Rings, Clusters, and Polymers of the Main Group Elements

Alan H. Cowley, EDITOR

University of Texas at Austin

Based on a symposium sponsored by the ACS Division of Inorganic Chemistry at the 184th Meeting of the American Chemical Society, Kansas City, Missouri, September 12–17, 1982

ACS SYMPOSIUM SERIES 232

AMERICAN CHEMICAL SOCIETY

WASHINGTON, D.C. 1983

Library of Congress Cataloging in Publication Data

Rings, clusters, and polymers of the main group elements. (ACS symposium series, ISSN 0097-6156; 232)

"Based on a symposium sponsored by the ACS Division of Inorganic Chemistry at the 184th Meeting of the American Chemical Society, Kansas City, Missouri, September 12-17, 1982."

Includes bibliographical references and index.

1. Ring formation (Chemistry)—Congresses.
2. Polymers and polymerization—Congresses.
3. Organometallic compounds—Congresses.

I. Cowley, Alan H. II. American Chemical Society. Division of Inorganic Chemistry. III. American Chemical Society. Meeting (184th: 1982: Kansas City, Mo.) IV. Series.

QD281.R5R56 1983 547'.5 83-15462
ISBN 0-8412-0801-8

PRINTED IN THE UNITED STATES OF AMERICA

ACS Symposium Series

M. Joan Comstock, *Series Editor*

FOREWORD

The ACS SYMPOSIUM SERIES was founded in 1974 to provide a medium for publishing symposia quickly in book form. The format of the Series parallels that of the continuing ADVANCES IN CHEMISTRY SERIES except that in order to save time the papers are not typeset but are reproduced as they are submitted by the authors in camera-ready form. Papers are reviewed under the supervision of the Editors with the assistance of the Series Advisory Board and are selected to maintain the integrity of the symposia; however, verbatim reproductions of previously published papers are not accepted. Both reviews and reports of research are acceptable since symposia may embrace both types of presentation.

CONTENTS

PREFACE

STUDY OF THE MAIN GROUP ELEMENTS has consistently promoted advances in many fields of chemistry. The mainstream of research, though, has been somewhat restricted in the past to the study of relatively small molecules. More recently, the increased knowledge of molecular topology, bonding, and coordination chemistry, coupled with safer and simpler syntheses of reasonable yield, has forged a path for more productive study of the larger ring, cluster, and polymer compounds of this group. Although the puzzles to be solved are still extremely complex and quite challenging, the relatively small group of researchers who tackle macromolecular studies perceive the promise of exciting and dramatic advances in the near future in both fundamental science and technology.

The Ralph Rudolph Memorial Symposium on rings, clusters, and polymers of the main group elements embraced a unique array of topics in this diverse field presented by a remarkable group of experts from around the world. Their contributions in this volume establish a long-needed definitive work in this now fast-moving area. The authors outline the realized and potential impact of the advances in their subspecialties—a very valuable perspective—and detail specific examples from their own research.

It is hoped that this unprecedented collection will increase interaction among researchers in this area of inorganic chemistry and researchers in organic chemistry, polymer chemistry, and applied material science; the rewards are undeniable.

IN MEMORIAM: Ralph W. Rudolph

Adapted from a speech by Robert W. Parry, President of the American Chemical Society, 1982

Ralph W. Rudolph, a brilliant and innovative inorganic chemist, was a proponent of active communication within his field and across related disciplines. Initially his research interests focused on fluorophosphines, but he continually explored new and challenging areas to the benefit of the inorganic community. He used his pioneering expertise to decipher mysteries of compounds having phosphorus–phosphorus bonds, polyboranes and carboranes, and complex cage structures involving various main group elements. He strove to devise classification systems much needed in this undefined area. He wrote, "speaking personally, my research has attracted much more attention and become more fulfilling as it has evolved and diversified from boron clusters to include main group metal clusters, catalysis, etc. Many have suggested that although specialization has some advantages, creativity usually fluorishes with diversification."

Not surprisingly, it was Ralph Rudolph who foresaw the great need for a comprehensive forum in the fast-moving area of rings, clusters, and polymers of the main group elements. The purposes of this symposium were to communicate advances to the inorganic community, to fill the gap for a much needed definitive work in this area, and to spur greater interaction with organic chemistry, transition metal chemistry, and applied science that is needed to fully develop this field. In response to his proposal for such a unifying symposium, the Inorganic Division requested that he organize the program. He agreed, and things were going well until late in April 1981. At that time, Ralph went into the hospital for a check-up. He had been suffering from Hodgkin's disease, but his progress seemed wonderful. Then, one week later he became dreadfully ill and passed away in Ann Arbor on May 11, 1981. Thanks to the efforts of Alan Cowley, the distinguished panel of speakers, and those who attended the symposium, the program envisioned by Ralph was a monumental success, a very fitting tribute to this bright and dedicated scientist.

Ralph Rudolph was born on July 14, 1940 in Erie, Pennsylvania. In 1962, he was graduated from Pennsylvania State University very near the top of his class. He received the Ph.D. from the University of Michigan in 1966. His thesis, which reported the synthesis of PF_2H, P_2F_4, and PF_2CN, initiated many studies, both theoretical and experimental, in laboratories around the world. Following his graduation from Michigan, he spent three years in Fort Collins, Colorado, at the F. J. Seiler Research Laboratory of the U.S. Air Force. In 1969, he returned to the University of Michigan as a member of the faculty. At the time of his death he was developing a program of research in the chemistry of "naked metal clusters."

He will be remembered for his pioneering syntheses in phosphorus chemistry, for his synthesis of thiaborane (a boron cage with a sulfur atom incorporated into the cage), for his inspirational teaching at both the graduate and undergraduate levels, for his strong sense of professional responsibility which is illustrated by the birth of this symposium, and for his many personal attributes which won him worldwide recognition. He was a delightful human being with very high standards of integrity and achievement.

Systematic Approaches to the Preparation of Boron Hydrides and Their Derivatives

S. G. SHORE

The Ohio State University, Department of Chemistry, Columbus, OH 43210

Classical procedures for boron hydride syntheses are reviewed. The development of principles of syntheses is presented and practical preparations of boron hydrides and their derivatives are described.

One of the main drawbacks to the development of the chemistry of many of the boron hydrides has been the absence of synthetic procedures for producing these materials in reasonable yields and quantities by relatively safe and simple techniques. Classical approaches are heavily dependent upon pyrolytic procedures. Although they have been developed to a "fine art," they require a high degree of skill in order to be employed safely in ordinary laboratory environments. Other important classical methods are dependent upon controlled protolysis reactions, frequently giving mixtures of materials which are difficult to separate.

For some time we have been concerned with developing rational approaches to boron hydride syntheses which would be based upon the systematic application of chemical principles. This approach has proved to be successful for the preparation of a number of boron hydrides. Described below is the development of principles of boron hydride syntheses and their applications.

Background

The first definitive preparation and characterization of boron hydrides was reported in 1912 (1). With this work, Alfred Stock and his collaborators introduced an area of inorganic chemistry for which there was no precedent in terms of structure and bonding. Over a period of twenty-four years this group isolated, identified and studied the chemistry of six boron hydrides (B_2H_6, B_4H_{10}, B_5H_9, B_5H_{11}, B_6H_{10}, and $B_{10}H_{14}$). In that particular time frame (1912-1936) these efforts represented remarkable achievements since it was first necessary to develop techniques for handling, separating and physically character-

0097-6156/83/0232-0001 $06.00/0

izing materials (frequently in less than millimolar quantity) which for the most part were volatile substances that react violently with air (2).

The earliest syntheses of boron hydrides were grossly inefficient, giving mixtures of products of 4-5% total yields. These procedures were based upon the reaction of magnesium boride with hydrochloric acid. Only much later were yields of 11% obtained when 8N phosphoric acid was used (2).

$$Mg_3B_2 + H_3PO_4 \longrightarrow \text{Boron Hydrides} + H_2 \qquad (1)$$

$$\left[\begin{array}{c} B_4H_{10},\ B_5H_9, \\ B_6H_{10},\ B_{10}H_{14}(\text{trace}) \end{array}\right]$$

Diborane(6) was obtained from the thermal decomposition of B_4H_{10} (2). In view of the restricted amounts of products obtained by the use of magnesium boride in boron hydride syntheses, development of their chemistry proceeded slowly even after techniques had been acquired for handling these materials.

The first significant improvement in boron hydride syntheses occurred in 1931 when Schlesinger and Burg (3) obtained B_2H_6 in overall yields of 75%, albeit in small amounts, by passing BCl_3 and H_2 through a high voltage discharge. Over a period of approximately ten years Schlesinger and Burg (4), along with a number of associates, continued to improve upon the preparation of B_2H_6, and they also pioneered in studying the chemistry of this hydride. This work in conjunction with the discovery of metal borohydrides (5-12) made possible later large scale preparations of B_2H_6.

Based upon a halide-hydride exchange reaction, the first truly practical synthesis of B_2H_6 was reported by Finholt, Bond, and Schlesinger in 1947 (13) (Reaction (2)).

$$3\ LiAlH_4 + 4BCl_3 \xrightarrow{Et_2O} 2B_2H_6 + 3LiCl + 3AlCl_3 \qquad (2)$$

Yields of 99% B_2H_6 were obtained. Subsequently, a number of viable procedures were developed for the preparation of B_2H_6 by the halide-hydride exchange route (14,15). Prominent among these is the following reaction (16):

$$3\ NaBH_4 + 4\ BF_3 \xrightarrow{\text{diglyme}} 2B_2H_6 + 3\ NaBF_4 \qquad (3)$$

Other useful routes to B_2H_6 were also delineated. Among these is the reaction of NaH with excess $B(OCH_3)_3$ (14,15,17), the reaction of BH_4^- with I_2 (18,19), and the controlled reaction of BH_4^- with H_3PO_4(conc) (20). In a later section of this article there is reported a high yield preparation of B_2H_6 by hydride abstraction from BH_4^- in the absence of a solvent (21).

Attempts to develop high energy borane fuels in the 1950's

elevated B_2H_6, B_5H_9, and $B_{10}H_{14}$ from the status of laboratory curiosities to compounds produced on a tonnage basis. From the controlled pyrolysis of B_2H_6, either B_5H_9 or $B_{10}H_{14}$ as the desired product, was obtained (22). This program spawned intense research activity, leading to rapid developments in synthetic chemistry (organoboranes (23), hydroboration (24,25), polyhedral boranes (26,27), carboranes (27,49), metalloboranes (27,50), metallocarboranes (27,50)) and structural and theoretical chemistry (27,28). Today, a number of B_2H_6 derivatives are commercially available, and $B_{10}H_{14}$ serves as the starting material for a large variety of higher boranes, carboranes, metalloboranes, and metallocarboranes. Uses for B_5H_9 have been primarily concerned with its application as a starting material in the preparation of smaller carboranes, metalloboranes, and metallocarboranes.

In the 1950's and 1960's a number of new boron hydrides and boron hydride anions were prepared. Although particular emphasis was placed upon stable higher borane entities derived from $B_{10}H_{14}$, significant progress was made in the preparation of new, smaller boron hydrides. Unfortunately, at this period of time, most of these materials could be obtained only in small amounts. (A comprehensive treatment of all boron hydrides prepared through 1978 is given in references (51) and (52).

During this period of progress, significant procedures for the already known lower boron hydrides B_4H_{10} (15,29,30) and B_5H_{11}(30,31) were introduced, but these posed technical demands and safety problems, which in general made them unattractive as laboratory preparations. Subsequent, relatively safe preparations of B_4H_{10} were achieved through the reaction of $[N(CH_3)_4][B_3H_8]$ with polyphosphoric acid (40% yield) (32) and NaB_3H_8 with HCl (40% yield) (33), but suffered from the requirement of time-consuming and tedious separation procedures.

By the close of the 1960's emphasis had shifted to the preparation and study of metalloboranes and metallocarboranes, many of which were prepared from $B_{10}H_{14}$ and B_5H_9 (legacies from the high energy fuels program). Interest in boron hydrides which could not be readily obtained or prepared had waned, not because they were intrinsically uninteresting, but more likely because of problems associated with obtaining them. At this time, from studies of B_5H_9, we became increasingly aware of possibilities for developing rational, systematic methods for preparing some of the rare smaller boron hydrides in good yields and reasonable quantities. This work eventually led to the development of two different types of systematic approaches to boron hydride syntheses (34,21). It not only resulted in the preparation of boron hydrides previously not readily accessible, but also recently demonstrated the potential of B_5H_9 as a feedstock in the preparation of $B_{10}H_{14}$ and other higher boron hydride systems and derivatives. Described below are the results of these studies.

Deprotonation and Borane Insertion Reactions

Many of the boron hydrides have been shown to function as Brönsted acids (See 34 for references). The bridging hydrogens are acidic. Pentaborane(9) is a monoprotic acid in the presence of a variety of bases (35,36,37). An example of such a reaction (38) is given in Scheme I, Reaction (4). By removing the bridging proton to give $B_5H_8^-$, a boron-boron bond is formed which is susceptible to insertion of electrophilic agents. Thus by inserting BH_3 into this site, the new anion $B_6H_{11}^-$ was formed (35,39) (Scheme I, Reaction (5)). Treatment of this anion with HCl gave B_6H_{12} in 60% yields (39) (Scheme I, Reaction (6)). On the other hand by treating $B_6H_{11}^-$ with an additional quantity of B_2H_6, B_6H_{10} was obtained in yields of 30% (40) (Scheme I, Reaction (7)).

The procedures outlined above demonstrate rational principles for preparing two boron hydrides previously obtained only with difficulty in very small yields as apparent by-products of reactions. The commercial availability of the starting material B_5H_9 (Callery Chemical Co., Callery, PA 16024) enhances the usefulness of these preparations.

This approach to boron hydride syntheses was further applied through the following sequence which started with B_4H_{10} and gave B_5H_{11} in 60% yields (39) (Scheme II, Reactions (8), (9), (10).

The preparation of B_6H_{10} was greatly improved when brominated pentaborane(9) was used as a starting material (41).

$$\text{1-BrB}_5\text{H}_8 + \text{KH} \longrightarrow \text{K[1-BrB}_5\text{H}_7] + \text{H}_2 \qquad (11)$$

$$\text{K[1-BrB}_5\text{H}_7] + {}^1\!/_2\,\text{B}_2\text{H}_6 \longrightarrow \text{B}_6\text{H}_{10} + \text{KBr} \qquad (12)$$

Yields of B_6H_{10} were 75%. This method demonstrates that bromide ion is a good leaving group from borane entities. In subsequent studies by Schaeffer (42) and by Gaines (43), fused borane and linked borane cages were prepared by taking advantage of this property.

Hydride Abstraction Reactions in the Absence of a Solvent

In the course of studying the preparation of B_6H_{10} from $B_6H_{11}^-$ (Reaction (7)), significant quantities of $B_{10}H_{14}$ (up to 27% yields) (40) were obtained from reaction mixtures in ethers maintained at or above room temperature. The source of $B_{10}H_{14}$ was believed to be $B_6H_{11}^-$ from which BH_4^- was apparently eliminated upon warming the system.

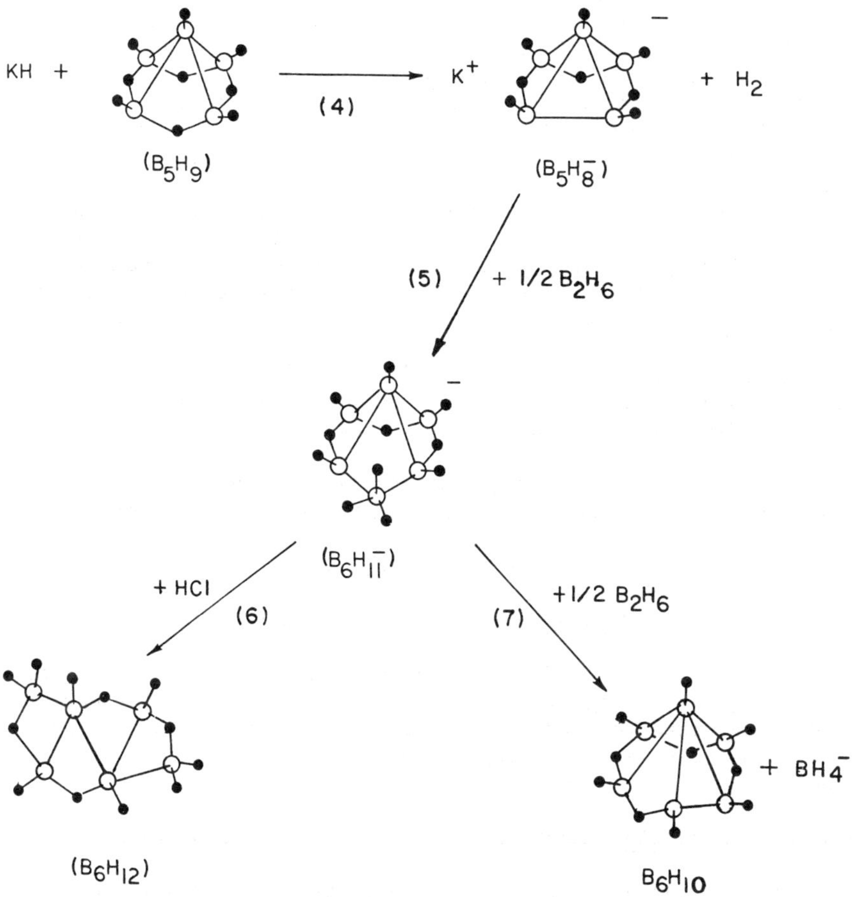
KH +
(4)
K^+
+ H_2
(B_5H_9)
$(B_5H_8^-)$
(5)
+ 1/2 B_2H_6
$(B_6H_{11}^-)$
+ HCl
(6)
(7)
+1/2 B_2H_6
(B_6H_{12})
B_6H_{10}
+ BH_4^-

SCHEME I

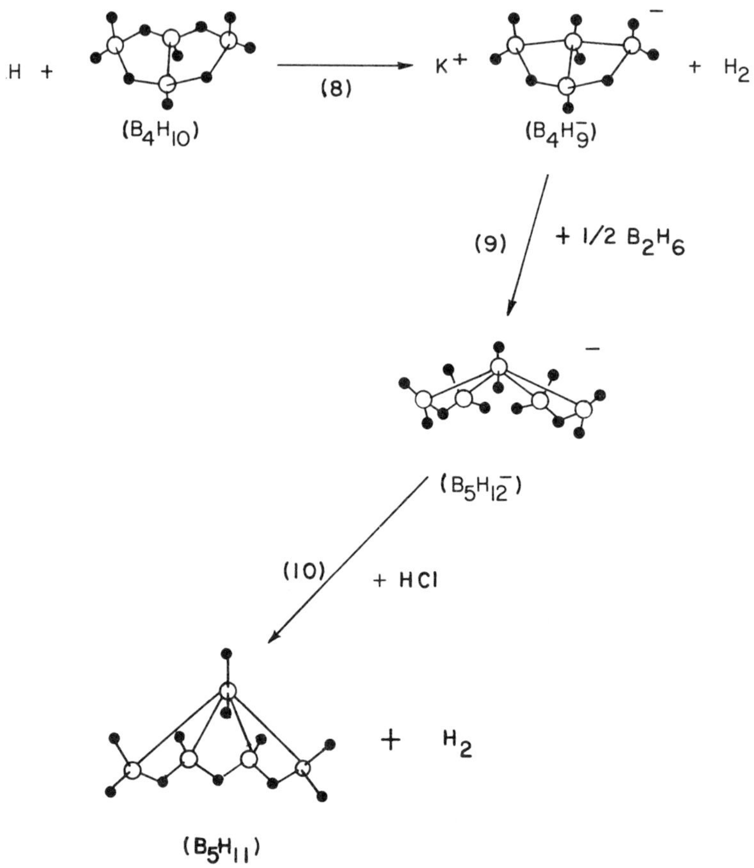

SCHEME II

$$K[B_5H_8] + \tfrac{1}{2} B_2H_6 \longrightarrow K[B_6H_{11}] \longrightarrow [B_5H_7] + K[BH_4] \quad (13)$$
$$[B_5H_7] \downarrow \tfrac{1}{2} B_{10}H_{14}$$

The possibility was considered that if Reaction (13) does occur through the intermediate $[B_5H_7]$, then $1\text{-}BrB_5H_7^-$ might be a very good candidate for preparing $B_{10}H_{14}$ in high yields, since Br^- is a good leaving group. The following reaction was attempted.

$$K[1\text{-}BrB_5H_7] + BBr_3 \longrightarrow [B_5H_7] + K[BBr_4] \quad (14)$$
$$[B_5H_7] \downarrow \tfrac{1}{2} B_{10}H_{14}$$

It was a failure. Although $B_{10}H_{14}$ was detected spectroscopically, the yield was insignificant and the reaction mixture was intractable.

A second attempt was made at bromide ion abstraction. In this case, however, the $B_3H_7Br^-$ ion was used in the hope of preparing B_6H_{12} through the following proposed reaction.

$$[N(\eta\text{-}C_4H_9)_4][B_3H_7Br] + BBr_3 \longrightarrow [B_3H_7] + [N(\eta\text{-}C_4H_9)_4][BBr_4]$$
$$[B_3H_7] \downarrow \tfrac{1}{2} B_6H_{14} \longrightarrow \tfrac{1}{2} B_6H_{12} + \tfrac{1}{2} H_2 \quad (15)$$

In the sense that B_6H_{12} was not produced, this reaction was also a failure; however it did produce $2\text{-}BrB_4H_9$ and B_4H_{10} (each in 10-15% yield). In fact, this is the preferred route to $2\text{-}BrB_4H_9$. Furthermore, this result suggested that the H^- ion as well as Br^- can be abstracted from $B_3H_7Br^-$. Thus if $B_3H_8^-$ were used in place of $BrB_3H_7^-$, the yield of B_4H_{10} should increase. Indeed this proved to be the case and provided the key to establishing a systematic route to boron hydride syntheses to produce B_2H_6, B_4H_{10}, B_5H_{11}, and $B_{10}H_{14}$ (21).

The procedures are simple and give essentially stoichiometric reactions which can be carried out in the absence of a solvent. The systematic nature of these syntheses relates to the observation that hydride ion can be abstracted from certain boron hydride anions to give as one of the final products a neutral boron hydride which contains one more boron atom than the anionic starting material. These reactions are described below.

B_2H_6 from BH_4^-. The simplest reaction observed involves the abstraction of hydride ion from BH_4^- by a boron halide to generate BH_3 units which combine to form B_2H_6. Reaction (16) represents the general reaction observed.

$$2\ BH_4^- + 2\ BX_3 \longrightarrow 2\ BH_3 + 2\ HBX_3^-$$

$$BH_3 + BH_3 \longrightarrow B_2H_6$$

$$2\ BH_4^- + 2\ BX_3 \longrightarrow B_2H_6 + 2\ HBX_3^- \qquad (16)$$

The reaction of solid $NaBH_4$ with BF_3 at room temperature gives B_2H_6 in 95% yield. This procedure differs from Reaction (3) in that hydride-halide exchange is not involved and the system is free of ether solvent.

B_4H_{10} from $B_3H_8^-$. Tetraborane(10) is obtained in 65% yield (based on boron in $B_3H_8^-$) from the reaction of solid $[N(n\text{-}C_4H_9)_4][B_3H_8]$ with BCl_3 at room temperature. This reaction has been studied in detail. It is analogous to Reaction (16) and is viewed in the following way.

An initial step involves abstraction of hydride ion from $B_3H_8^-$ to generate the unstable intermediate $[B_3H_7]$. Transfer of a BH_3 unit from one $[B_3H_7]$ molecule to another would generate B_4H_{10} and a residue of empirical composition $(BH_2)_x$. Thus the following sequence with resulting stoichiometry (17) is envisioned.

$$B_3H_8^- + BCl_3 \longrightarrow [B_3H_7] + HBCl_3^-$$

$$\tfrac{1}{2}[B_3H_7] + \tfrac{1}{2}[B_3H_7] \longrightarrow \tfrac{1}{2}B_4H_{10} + 1/x(BH_2)_x$$

$$B_3H_8^- + BCl_3 \longrightarrow \tfrac{1}{2}B_4H_{10} + HBCl_3^- + 1/x(BH_2)_x \qquad (17)$$

In reaction (17), 67% of the boron in $B_3H_8^-$ is converted to B_4H_{10}. This percent conversion agrees closely with experimental yields. It suggests that the preparative procedure is quantitative with respect to the theoretical amount of B_4H_{10} which can be produced. This is by far the safest route to B_4H_{10} available. It is also attractive because of the commercial availability of $[N(CH_3)_4][B_3H_8]$ (Alfa Products, Danvers, MA 01923).

B_5H_{11} from $B_4H_9^-$. Pentaborane(11) is obtained in 60% yield (based upon boron in $B_4H_9^-$) from the reaction of solid $K[B_4H_9]$ with liquid BCl_3 at $-35^\circ C$. This reaction has been studied in detail. It is analogous to reactions (16) and (17) and is viewed as follows.

Hydride abstraction from $B_4H_9^-$ generates the unstable intermediate $[B_4H_8]$. Formation of B_5H_{11} can occur in a second step which involves transfer of BH_3 from one $[B_4H_8]$ unit to another $[B_4H_8]$. The following sequence with resulting stoichiometry (18) is suggested.

$$B_4H_9^- + BCl_3 \longrightarrow [B_4H_8] + HBCl_3^-$$
$$\tfrac{1}{2}[B_4H_8] + \tfrac{1}{2}[B_4H_8] \longrightarrow \tfrac{1}{2}B_5H_{11} + (\tfrac{1}{2}x)(B_3H_5)_x$$

$$B_4H_9^- + BCl_3 \longrightarrow \tfrac{1}{2}B_5H_{11} + HBCl_3^- + (\tfrac{1}{2}x)(B_3H_5)_x \qquad (18)$$

In reaction (18) 63% of the boron in the $B_4H_9^-$ is converted to B_5H_{11}. Thus the experimental yield, 60%, is close to the theoretical limit of B_5H_{11} available from (18). This is the safest route to B_5H_{11}.

$B_{10}H_{14}$ from $B_9H_{14}^-$. Decaborane(14) is obtained in 50% yield (based upon boron in $B_9H_{14}^-$) from the reaction of solid $[N(CH_3)_4][B_9H_{14}]$ with BCl_3 at room temperature. This reaction has been studied in detail. It is analogous to Reactions (16), (17), and (18) and is believed to occur through the following sequence giving stoichiometry (19).

$$B_9H_{14}^- + BCl_3 \longrightarrow [B_9H_{13}] + HBCl_3^-$$
$$\tfrac{1}{2}B_9H_{13} + \tfrac{1}{2}B_9H_{13} \longrightarrow \tfrac{1}{2}B_{10}H_{14} + \tfrac{1}{2}x(B_8H_{10})_x + \tfrac{1}{2}H_2$$

$$B_9H_{14}^- + BCl_3 \longrightarrow \tfrac{1}{2}B_{10}H_{14} + \tfrac{1}{2}H_2 + HBCl_3^- + (\tfrac{1}{2}x)(B_8H_{10})_x \qquad (19)$$

In reaction (19) 56% of the boron in the $B_9H_{14}^-$ is converted to $B_{10}H_{14}$. The experimental yield of $B_{10}H_{14}$, 50%, suggests that the theoretical limit defined by Reaction (19) is approached.

In the above preparations of $B_{10}H_{14}$, B_5H_{11}, and B_4H_{10}, abstraction of H^- from a borane anion gives an unstable neutral fragment which is believed to obtain a BH_3 unit from another like fragment to form the desired borane product. One hundred per cent conversion of the starting material to the desired boron hydride cannot be achieved. On the other hand if BH_3 could be "funneled" into the reaction mixture to react with the unstable intermediate, then 100% conversion could be achieved in principle.

The ability of $ClBH_3^-$ to function as a BH_3 donor was examined in the preparation of $B_{10}H_{14}$. An equivalent of $[N(CH_3)_4][ClBH_3]$ was added to $[N(CH_3)_4][B_9H_{14}]$ along with double the amount of BCl_3 normally used to prepare $B_{10}H_{14}$ The purpose of the excess BCl_3 was to extract Cl^- from $ClBH_3^-$, thereby releasing BH_3 to react with the $[B_9H_{13}]$ generated from the reaction of $B_9H_{14}^-$ with BCl_3. The projected stoichiometry is Reaction (20), with complete conversion of $B_9H_{14}^-$ to $B_{10}H_{14}$.

$$B_9H_{14}^- + ClBH_3^- + 2BCl_3 \longrightarrow B_{10}H_{14} + H_2 + HBCl_3^- + BCl_4^- \qquad (20)$$

A yield of 70% $B_{10}H_{14}$ (based upon the boron in B_9H_{14} and $ClBH_3^-$) was obtained. This result supports the proposed sequence for

Reaction (19) and presents a promising modification of the synthetic procedure.

Pentaborane(9): A Raw Material for the Preparation of Boron Hydrides and Boron Hydride Derivatives.

Pentaborane(9) is a potentially useful raw material for the production of higher boron hydrides. It is commercially available (Callery Chemical Co.; Callery, PA 16024). Close to 100 tons is in storage, left over from the high energy borane fuels program. Its use in the preparation of B_6H_{10} and B_6H_{12} has already been discussed in this article. Other, perhaps more important applications can be made also. The examples given below represent a beginning in the study of such applications.

$B_{10}H_{14}$ from B_5H_9 via $B_9H_{14}^-$. A convenient, "one pot" procedure has been developed for the preparation of $B_{10}H_{14}$ from B_5H_9 which is amenable to a wide range of reaction scales (21). In one step the B_5H_9 is converted to $[N(CH_3)_4][B_9H_{14}]$, and in the second step $B_9H_{14}^-$ is converted to $B_{10}H_{14}$ according to the preparation discussed previously (See Scheme III). The $B_{10}H_{14}$ is then sublimed from the reactor.

Conversion of B_5H_9 to $[N(CH_3)_4][B_9H_{14}]$ is achieved through the reaction of B_5H_9 with NaH in 2:1 molar ratio in THF at room temperature in the presence of one equivalent of $[N(CH_3)_4]Cl$. An initial deprotonation step is involved, as in Reaction (4), to generate $B_5H_8^-$ which then reacts with an equivalent amount of the remaining B_5H_9 to give $B_9H_{14}^-$.

$B_9H_{13}L$ and n-$B_{18}H_{22}$ from $B_9H_{14}^-$. Standard syntheses for $B_9H_{14}^-$, $B_9H_{13}L$ (L = Lewis base) and n-$B_{18}H_{22}$ use $B_{10}H_{14}$ as a starting point and involve degradation of the B_{10} skeleton to a B_9 skeleton. By taking advantage of the simple-high yield conversion of B_5H_9 to $B_9H_{14}^-$, described above, it has been possible to convert B_5H_9 to $B_9H_{13}L$ (L = R_2S, NR_3, Ph_3P) and n-$B_{18}H_{22}$ in practical "one-pot" procedures which are modifications of earlier methods(44). The $B_9H_{14}^-$ is converted to $B_9H_{13}L$ and n-$B_{18}H_{22}$ according to Scheme IV below.

5,6-$(CH_3)_2C_2B_8H_{10}$ from B_5H_9 via $B_9H_{14}^-$. The carborane 5,6-$(CH_3)_2C_2B_8H_{10}$ was produced in 35% yield (based on boron in B_5H_9). Scheme V outlines the synthesis. This type of carborane was previously prepared from a larger carborane system through a degradation process (45,46) and also from the reaction of B_9H_{12} with alkynes (47,48). The procedure given in Scheme V is simpler than these earlier preparative methods.

Acknowledgments are made to the Army Research Office and the National Science Foundation for financial support.

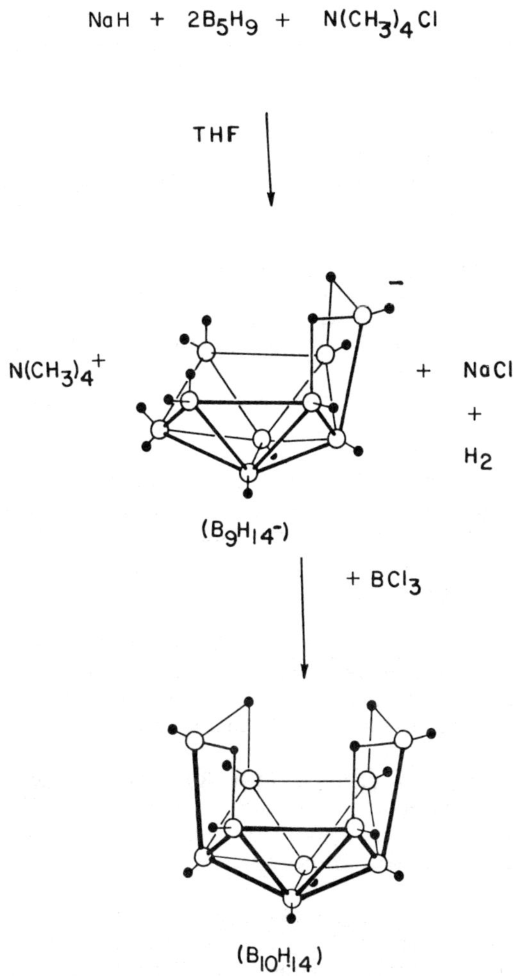

SCHEME III

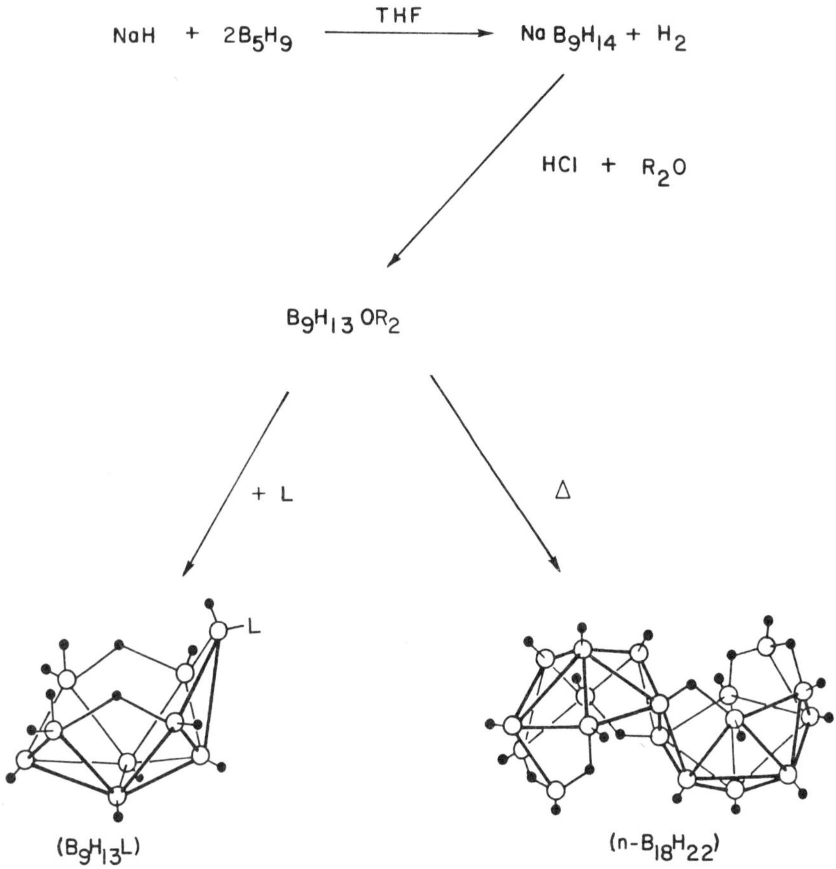
NaH + 2B5H9
THF
Na B9H14 + H2
HCl + R2O
B9H13 OR2
+ L
Δ
L
(B9H13L)
(n-B18H22)

SCHEME IV

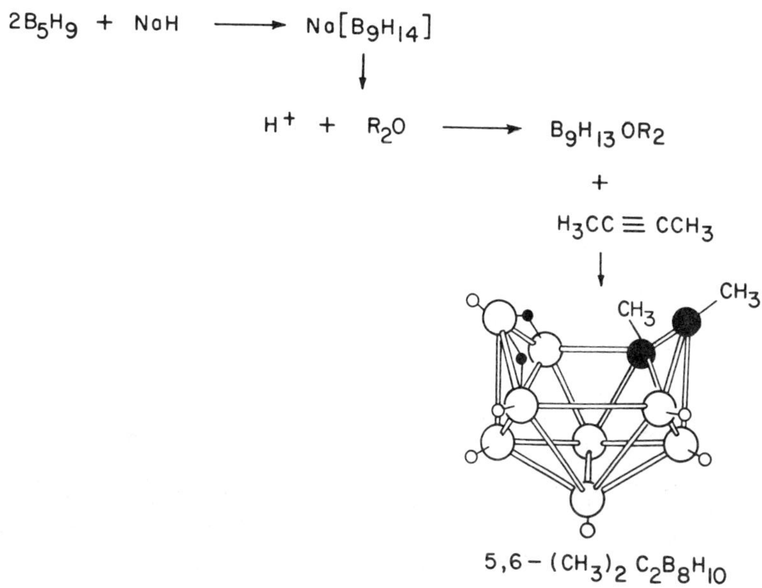

$2B_5H_9$ + NaH → $Na[B_9H_{14}]$
H^+ + R_2O → $B_9H_{13}OR_2$
+
$H_3CC \equiv CCH_3$
CH_3
CH_3
$5,6-(CH_3)_2C_2B_8H_{10}$

SCHEME V

Literature Cited

1. Stock, A; Massenez, C. Ber. 1912, 45, 3529.
2. Stock, A. "Hydrides of Boron and Silicon," Cornell University Press, Ithaca, New York, 1933.
3. Schlesinger, H. I.; Burg, A. B. J. Am. Chem. Soc. 1931, 53, 4321.
4. Schlesinger, H. I..; Burg, A. B. Chem. Rev. 1942, 31, 1.
5. Schlesinger, H. I. Brown, H. C. J. Am. Chem. Soc. 1940, 62, 3429.
6. Burg, A. B.; Schlesinger, H. I. J. Am. Chem. Soc. 1940, 62, 3425.
7. Schlesinger, H. I.; Sanderson, R. T.; Burg, A. B. J. Am. Chem. Soc. 1939, 61, 536.
8. Schlesinger, H. I.; Sanderson, R. T.; Burg, A. B. J. Am. Chem. Soc. 1940, 62, 3421.
9. Schlesinger, H. I.; Brown, H. C.; Abraham, B.; Bond, A. C.; Davidson, N.; Finholt, A. E.; Gilbreath, J. R.; Hoekstra, H.; Horvitz, L.; Hyde, E. K.; Katz, J. J.; Knight, J.; Lad, R. A.; Mayfield, D. L.; Rapp, L.; Ritter, D. M.; Schwartz, A. M.; Sheft, I.; Tuck, L. D.; and Walker, A. O. J. Am. Chem. Soc. 1953, 75, 186.
10. Schlesinger, H. I.; Brown, H. C.; Hoekstra, H. R.; Rapp, L. J. Am. Chem. Soc. 1953, 75, 199.
11. Schlesinger, H. I.; Brown, H. C., Finholt, A. E. J. Am. Chem. Soc. 1953, 75, 205.
12. Schlesinger, H. I.; Brown, H. C.; Hyde, E. K. J. Am. Chem. Soc. 1953, 75, 209.
13. Finholt, A. E.; Bond, A. C.; Schlesinger, H. I. J. Am. Chem. Soc. 1947, 69, 1199.
14. Adams, R. M. Advan. Chem. Ser. 1961, 33, 60.
15. Parry, R. W.; Walter, M. K. Prep. Inorg. React. 1968, 5, 45.
16. Brown, H. C.; Tierney, P. J. J. Am. Chem. Soc. 1958, 80, 1522.
17. Bush, J. D.; Carpenter, R. A.; Schechter, W. H., U. S. Pat. 3,014,059; Dec. 19, 1961.
18. Nainan, K. C.; Ryschkewitsch Inorg. Nucl. Chem. Lett. 1970, 6, 765.
19. Freeguard, G. F.; Long, L. M. Chem. Ind. (London) 1965, 11, 471.
20. Duke, B. J.; Gilbert, J. R.; Read, I. A. J. Chem. Soc., London 1964, 50,
21. Toft, M. A.; Leach, J. B.; Himpsl, F. L.; Shore, S. G. Inorg. Chem. 1982, 21, 1952.
22. Holzman, R. T.; Hughes, R. L.; Smith, I. C.; Lawless, E. W., "Production of the Boranes and Related Research," Academic Press, New York, 1967.
23. Onak, T. "Organoboranes," Academic Press, New York, 1975.
24. Brown, H. C., "Hydroboration," Benajmin, New York, 1962.

25. Brown, H. C. "Boranes in Organic Chemistry," Cornell University Press, Ithaca, New York, 1972.
26. Muetterties, E. L.; Knoth, W. H. "Polyhedral Boranes," Dekker, New York, 1968.
27. Muetterties, E. L., "Boron Hydride Chemistry," Academic Press, New York, 1975.
28. Lipscomb, W. N. "Boron Hydrides," Benjamin, New York, 1963.
29. Kodama, G., Ph.D., Thesis, University of Michigan, Ann Arbor, 1957, pp. 60-62.
30. Klein, M. J.; Harrison, B. C.; Solomon, I. J. J. Am. Chem. Soc. 1958, 80, 4149.
31. Norman, A. D.; Schaeffer, R. Inorg. Chem. 1965, 4, 1225.
32. Gaines, D. F.; Schaeffer, R. Inorg. Chem. 1964, 3, 438.
33. Schaeffer, R.; Tebbe, F. J. Am. Chem. Soc. 1962, 84, 3974.
34. Shore, S. G. Pure and Appl. Chem. 1977, 49, 717.
35. Geanangel, R. A.; Shore, S. G. J. Am. Chem. Soc. 1967, 89, 6771.
36. Gaines, D. F.; Irons, T. V. J. Am. Chem. Soc. 1967, 89, 3375.
37. Onak, T.; Dunks, G. B.; Searcy, I. W.; Spielman, J. Inorg. Chem. 1967, 6, 1465.
38. Geanangel, R. A.; Johnson, H. D. II; Shore, S. G. Inorg. Chem. 1970, 9, 908.
39. Remmel, R. J.; Johnson, H. D., II; Jaworiwsky, I. S.; Shore, S. G. J. Am. Chem. Soc. 1975, 97, 5395.
40. Geanangel, R. A.; Johnson, H. D., II; Shore, S. G. Inorganic. Chem. 1971, 10, 2363.
41. Johnson, H. D., II; Brice, V.T.,; Shore, S. G. Inorg. Chem. 1973, 12, 689.
42. Huffman, J. C.; Moody, D. C.; Schaeffer, R. Inorg. Chem. 1976, 15, 227.
43. Gaines, D. F.; Jorgenson, M. W.; Kulzick, M. A. J. Chem. Soc., Chem. Commun. 1979, 380.
44. Dobson, J.; Keller, P. C.; Schaeffer, R. Inorgan. Chem. 1968, 7, 399.
45. Stibr, B.; Plesek, J.; Hermanek, S. Coll. Czech. Chem. Commun. 1873, 38, 338.
46. Plesek, J.; Hermanek, S. Coll. Czech. Chem. Commun. 1974, 39, 821.
47. Rietz, R. R.; Schaeffer, R. J. Am. Chem. Soc. 1971, 93, 1263.
48. Rietz, R. R.; Schaeffer, R. J. Am. Chem. Soc. 1973, 95, 6254.
49. Grimes, R. "Carboranes," Academic Press, New York, 1970.
50. Grimes, R., "Metal Interactions with Boron Clusters," Plenum Press, New York, 1982.
51. Yamauchi, M.; Shore, S. G., "Gmelin Handbuch der Anorganischen Chemie," Springer-Verlag, Berlin, 1978, Boron Supplement 18, Vol. 52.

52. Barton, L., Onak, T., Shore, S. G., "Gmelin Handbuch der Anorganischen Chemie," Springer-Verlag, Berlin, 1979, Boron Supplement 20, Vol. 54.

RECEIVED June 21, 1983

2

Coordination and Related Chemistry of Polycyclic Tetraphosphorus Compounds

JEAN G. RIESS

Laboratoire de Chimie Minérale Moléculaire, Equipe de Recherche Associée au CNRS, Université de Nice, 06034 Nice, France

Polycyclic tetraphosphorus compounds, including the P_4 molecule, its oxides, sulfides, imides and mixed derivatives, and related compounds, offer a wide choice of potential donor sites. Their coordination chemistry is described and discussed. The transmission of structural and electronic effects through the molecular frame -i.e. the interdependence of the phosphorus donor sites - is evident in their chemical behavior, kinetic and thermodynamic differences, spectral and structural data. The interpretation of NMR and X-ray data in terms of changes in electronic distribution and of π character in bonding, is critically scrutinized. It is concluded that apportioning the observed effects among σ and π electronic contributions and steric contributions can usually not be achieved in a reliable way. Structural "softness" and disorder in the solid are recognized as prominent features characteristic of tetraphosphorus *closo*-compounds. The field contains many riddles, and promises new developments, especially in the realm of mixed picnogen/transition metal complexes and clusters.

Phosphorus displays a remarkable propensity to breed stable tetraphosphorus *closo*-type structures in many of its simplest molecular combinations, including its oxides, sulfides and imides (1-6). All derive, at least formally, from the P_4 tetrahedron. The oxides $(P_4O_6)O_n$ (n = 0 to 4), the related imides $[P_4(NR)_6](NR)_n$ (n = 0 to 4), and some of the sulfides, adopt the adamantanelike bonding arrangement $(A_4B_6)C_n$ (n = 0 to 4) shown in scheme 1. This arrangement is also found in an increasing range of mixed compounds that derive from the former by diverse substitutions in the branching, bridging and terminal positions (scheme 1); some of these were not isolated, but have only been identified in redistribution

0097–6156/83/0232–0017 $09.00/0

mixtures. The lower phosphorus sulfides, for their part, form a unique series of structures, generally of lesser symmetry, a sampling of which is shown in scheme 2.

All these compounds are close to spherical in shape (several of them show plastic-phase behavior), and are literally coated with valence-shell electron pairs, which make them potential multidentate donors. Although many of them have been known for over a century, it was not until 1965 that the first coordination adduct of a *closo*-phosphorus structure, $P_4O_6[(Ni(CO)_3]_4$ **1** (Figure 1) was reported (28).

Several books and general reviews are already available on phosphorus ring and cage* systems (16). This one will focus more specifically on the coordination chemistry of molecular tetraphosphorus-based *closo*-compounds, including P_4 itself and closely related analogs, and will also include structural data on both free and coordinated species. The questions of the transmission of electronic and structural effects through the molecular frame, and of the extent of π-bonding, will be critically considered. Some unpublished preliminary observations, as well as unresolved enigmas, will also be treated.

P_4 and P_4-related Tetrahedral Clusters as Donors.

Theoretical investigations of the P_4 molecule (30,31) depict its valence shell electron density as being concentrated inside and on the faces of the tetrahedron, where it plays the role of multicenter molecular bonding, with little or no (nucleophilic) free electron pairs located on, and pointing out from, the individual atoms.

At the same time the third-row-element based molecule is unsatured and expected to display π-acid character. Thus the most propitious conditions for bonding to a transition metal should be those under which the metal can both accept and back-donate electron density to the P_4 ligand.

The first transition metal adducts containing the tetraphosphorus molecule were reported by Ginsberg in 1971 (32). They resulted indeed from direct substitution of white phosphorus for phosphane or arsane ligands on low-valent rhodium :

$$L_3RhCl + P_4 \xrightarrow[78°]{} \underset{\mathbf{2}}{L_2RhCl(P_4)} + L$$

$$L = PPh_3,\ P(m\text{-tolyl})_3,\ P(p\text{-tolyl})_3,\ AsPh_3$$

The presence of the P_4 molecule was indicated by osmometric molecular weight determinations, mass spectrometry, and by the easy displacement, even at -78°C, of the intact P_4 unit by carbon monoxide or phosphane ligands. The infra-red and Raman spectra showed bands attributable to a bound P_4 molecule under C_{3v} or C_s

* a misleading denomination since there is no room in the "cage".

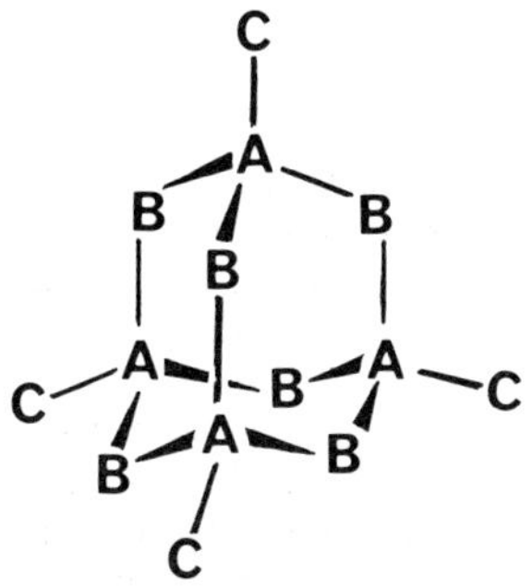

A : BRANCHING SITES
P, As, P/N, P/CR, As/CR

B : BRIDGING SITES
O, NR[a], S[b], Se
PPh[c], $SiMe_2$[c], $GeMe_2$[c], $SnMe_2$[c]
S/N^-[d], O/CH_2[e]

C : TERMINAL SITES
O, NPh, S, Se
O/S[f], O/Se[f]

Scheme 1. Examples of tetraphosphorus- or arsenic-based molecules with the adamantane-type core A_4B_6. Lewis acid/base complexes are not shown. Key to superscripts (when no reference is given, consult Ref. 1–6): [a]A unique non-adamantane $P_4(NR)_6$ closo-*structure has recently been obtained with $R = i\text{-}C_3H_7$ (7) and has been shown to convert into the thermodynamically more stable adamantane-type isomer upon heating; [b]as in P_4S_8 (8), P_4S_9, and P_4S_{10}; [c]Ref. 9; [d]in $P_4S_5N^-$ (10); [e]in $[P_4O_4(CH_2)_2]O_4$ (11) or $As_4O(CH_2)_2$ (12); [f]as in $(P_4O_6)O_{4-n}S_n$ (13) and $[P_4(NMe)_6]O_{4-n}S_n$ (14).*

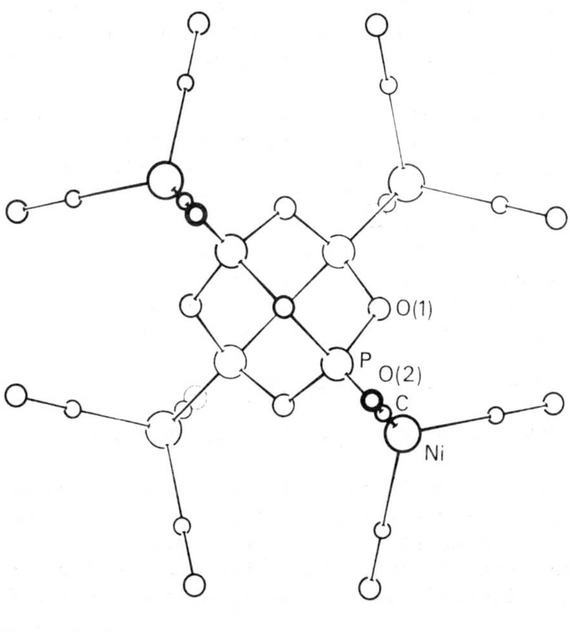

Figure 1. Structure of the $(P_4O_6)[Ni(CO)_3]_4$ adduct (29).

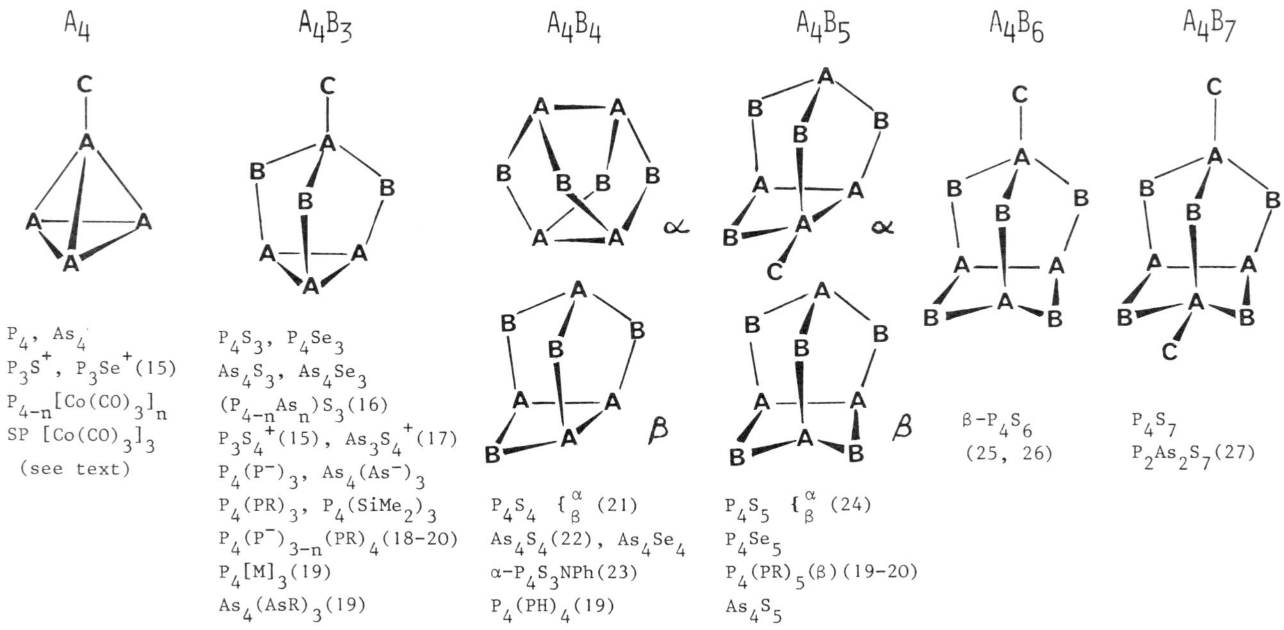

Scheme 2. Sampling of non-adamantane tetraphosphorus and related closo-*compounds* $A_4B_{0-6}C_{0-4}$.

symmetry but the mode of coordination of the P_4 molecule could not be determined exactly. In view of some of the above theoretical considerations, the hypothesis of bonding through a face of the tetrahedron was preferred over those through an edge or an apex. The failure to observe ^{31}P NMR lines due to P_4 suggested that some fast interor intramolecular exchange process was taking place on the NMR time scale.

Six years later the discrete diamagnetic complex $P_4[Fe(CO)_4]_3$ was obtained by Schmid and Kempney by allowing P_4 to react with $Fe_2(CO)_9$ (33). Its infra-red spectrum in solution indicated the presence of distinct $Fe(CO)_4$ groups, and the Mossbauer spectrum definitely showed the existence of penta- and hexa-coordinated iron atoms in a 1:2 ratio. The ^{31}P NMR, however, exhibited only a single line at 21 ppm at room temperature, which caused the authors to suggest a fluctuating structure in which two bridging and one apex-bound $Fe(CO)_4$ group exchange coordination sites rapidly in solution (Figure 2). The adduct was extremely air-, light- and temperature-sensitive, and decomposed within hours in benzene, toluene or THF solution, with evolution of CO and formation of a polymer $[(CO)_3FeO_2]_n$, which no longer contained the P_4 unit.

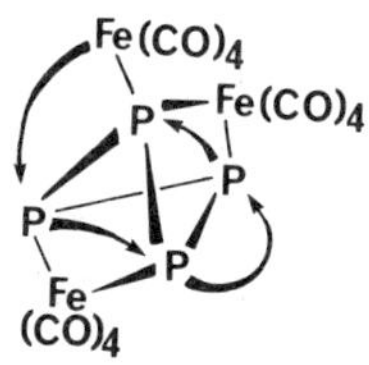

Figure 2. Proposed dynamic structure of $P_4[Fe(CO)_4]_3$ (33).

The nickel-P_4 and palladium-P_4 adducts **3** were prepared by Sacconi et al., by the following reaction (34) :

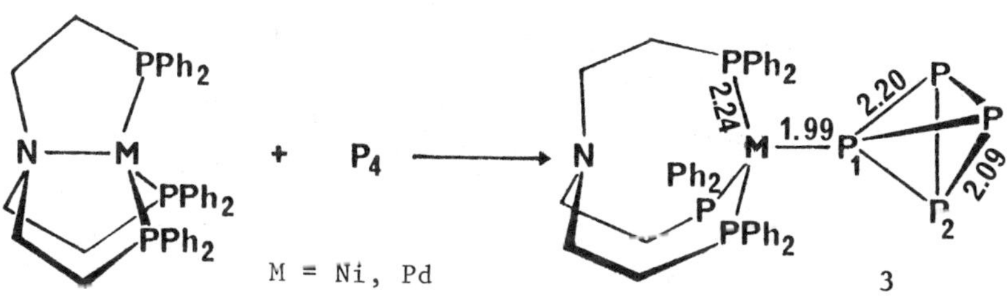

3

The nickel complex was stable enough to allow, for the first time, an X-ray diffraction analysis, which showed that the P_4 tetrahedron is bound to the metal through an apex. The nickel atom is surrounded by its four phosphorus ligands in a quite regular tetrahedral fashion, the nitrogen atom no longer being bound to the metal, i.e. reaction 2 can be described as a simple ligand exchange, with the P_4 molecule providing the metal with the two electrons initially donated by the nitrogen atom. The Ni-(P_4) bond is the shortest (by 0.25Å) of the four Ni-P bonds, suggesting a significant d_{π}-d_{π} interaction between the two atoms, and/or lower steric requirements for the P_4 molecule.

A distinct situation is created when reagents that are potential sources of a group based on a 15-electron transition metal are allowed to react with P_4. Such groups are electronically equivalent to phosphorus, in the sense that both the transition metal-based group and the pnicnogen element need 3 electrons in order to reach their stable closed-shell configurations of 18 and 8 electrons respectively (35, 36). One way to satisfy this need is, in either case, for the deficient center to share electrons with three other electronically analogous groups or elements; hence one has the analogy between the stable P_4 molecule and the $[Ir(CO)_3]_4$ cluster. This concept can be used to rationalize the existence of a series of fairly stable mixed $P_{4-n}[M]_n$ or $As_{4-n}[M]_n$ tetrahedral clusters obtained by Dahl (35,37), Marko (38,40), Sacconi (41,43) and co-workers, which can then be regarded as deriving from P_4 or As_4 by stepwise replacement of phosphorus or arsenic atoms by electronically equivalent metallic groups, for example (38) :

$$P_4 + Co_2(CO)_8 \longrightarrow \left\{ \begin{array}{l} P_3Co(CO)_3 \\ P_2[Co(CO)_3]_2 \\ P[Co(CO)_3]_3 \\ [PCo_3(CO)_8]_3 \end{array} \right.$$

The same series of products was also obtained by the reaction of PCl_3 or PBr_3 with $Co_2(CO)_8$ or that of PI_3 with $Co(CO)_4^-$ (38). Arsenic analogs had already been reported (35,37) by Dahl.

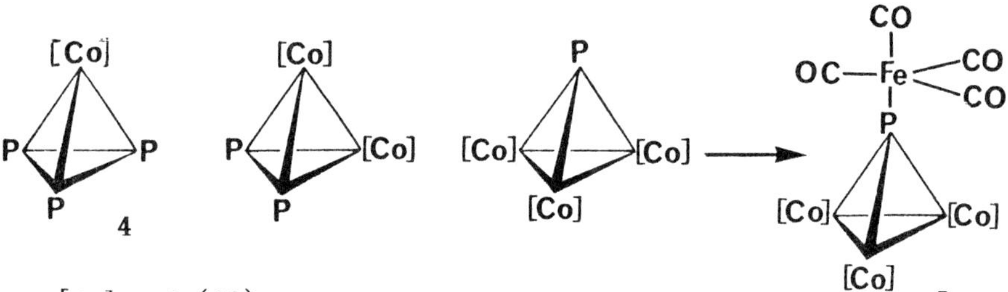

4

5

[Co] = $Co(CO)_3$

That these P_4-related tetrahedral clusters retain donor capacity through the phosphorus atoms was illustrated by the isolation of adduct 5 (the parent $P[Co(CO)_3]_3$ cluster is unstable (38), in which the PCo_3 cluster behaves as a two-electron donor towards the unsaturated 16-electron $Fe(CO)_4$ moiety. The rather high CO vibration frequencies in the infra-red spectrum indicate again that the cluster also has some π-accepting capacity. Further manifestation of the donor character of this phosphorus atom can be seen in its tendency to readily displace CO from a $Co(CO)_3$ moiety to form the remarkable cyclic trimer $\{E_3[Co(CO)_3]_2[Co(CO)_2]\}_3$ (E = P or As) (40), which results from the displacement of one carbonyl ligand on Co by the phosphorus apex of the next cluster in the cyclic trimer. $As[Co(CO)_3]_3$ readily gave the $M(CO)_5$ (M = Cr, Mo, W) and $AlCl_3$ adducts (unstable in the latter case) (41), and similar derivatives of the antimony cluster $Sb[\eta^5CpFe(CO)_2]_3$ have also been reported (42). The action of Lewis acids (or bases) on the $E_n[Co(CO)_3]_{4-n}$ clusters can however also provoke the redistribution of fragments within the cluster's core (43). For example :

$$As_3Co(CO)_3 + Co_2(CO)_8 \xrightarrow[W(CO)_6]{h\nu} As_2[Co(CO)_3]_2 + (CO)_5WAs[Co(CO)_3]_3 + \{As[Co(CO)_3]_2[Co(CO)_2]\}_3$$

In the same vein one must mention the recently reported hybrid clusters **6** :

$$RCl_2CLi + ECl_3 \longrightarrow RCl_2C\text{-}ECl_2 \xrightarrow{Co_2(CO)_8}$$

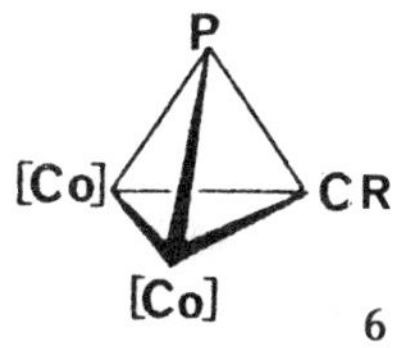

6

[Co] = $Co(CO)_3$; E = P, As; R = H (As only), Me, Ph, $SiMe_3$

which incorporate the three distinct electronically equivalent groups P, RC and $Co(CO)_3$ (44). These clusters may also be regarded as $Co_2(CO)_6$ adducts of the usually unstable phospha and arsaacetylene RC=E ligands, and are therefore related to the tetrahedral acetylene complexes $(RC{\equiv}CR)[(CO)_3Co\text{-}Co(CO)_3]$. The donor ability of the P (but not the As) atom is established by the formation of $(CO)_5MP(CR)[Co(CO)_3]_2$ adducts (M = Cr, Mo, W) when **6** is allowed to react with $M(CO)_5(thf)$, as well as by their slow cyclotrimerisation to $\{(MeC)P[Co(CO)_3][Co(CO)_2]\}_3$ (44).

The tetrahedral P_3ML clusters **7**, where M = Co, Ir, Rh, Pd or Pt, and L is a tripod ligand, obtained by Sacconi et al (34,45,47,49) :

$Co(H_2O)_6^{++}$ 2 BF_4^-
or $[Rh(CO)_2Cl]_2$
or $[Rh(C_2H_4)_2Cl]_2$
or $Ir(PPh_3)_2(CO)Cl$

$$\xrightarrow[CH_3C(CH_2PPh_2)_3]{P_4}$$

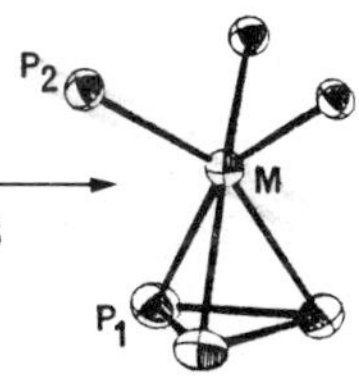

7

can be regarded, for the purposes of electron-counting, either as complexes in which the triphosphorus cycle acts as a tri-hapto 3-electron donor towards the transition metal moiety, or as tetrahedral P_3M clusters whose existence is based on 2-electron covalent bond formation. X-ray diffraction studies showed these compounds to be isomorphous (47). All contain the cyclotriphosphorus moiety and the triphosphorus tripod ligand, and all show an octahedral arrangement of the six phosphorus atoms around the metal. The $M(P_3)$ bonds are consistently longer than the other metal-P bonds, by 0.10 to 0.16 Å, while the P-P distances in the cyclo-P_3 unit are significantly shorter (2.14 to 2.16 Å) than in P_4 (2.21Å) or in the $P_3[Mn(CO)_2Cp]_3$ adduct (2.26 Å) (48). This shortening, as earlier proposed for the $As_3Co(CO)_3$ cluster (35), has been rationalized on the basis of partial delocalization of the electron density from the polarizable cyclo-P_3 fragment to the more electron-demanding $Co(CO)_3$ moiety, and consequent relief of interelectronic repulsions in the cyclo-P_3 fragment. In compensation the metal drains less electron density from the tripod's P atoms, which display longer Co-P bonds (by 0.10 to 0.15 Å) than usual. The slight increase in P-P bond-lengths, and concomitant decrease of the PMP angle, from cobalt to rhodium or iridium, is consistent with the increasing size of the metal atom and consequent reduction of orbital overlap with phosphorus.

That these mixed P_3M clusters still have electron-donor capacity is illustrated by :

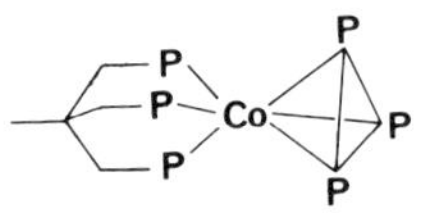

$$\xrightarrow{Cr(CO)_6} \begin{cases} (triphos)Co(\eta^3\text{-}P_3)Cr(CO)_5 \\ (triphos)Co(\eta^3\text{-}P_3)[Cr(CO)_5]_2 \end{cases}$$

8

in which the phosphorus atoms readily replace CO groups from $Cr(CO)_6$ to give stable adducts (49). The X-ray structure analysis performed on **8** showed (Figure 3) the two $Cr(CO)_5$ groups to be bound to individual phosphorus atoms, with Cr-P distances of 2.42 Å, equal to that found in $Ph_3P[Cr(CO)_5]$, suggesting that the Lewis base properties of the phosphorus atoms in the CoP_3 unit are similar to those of PPh_3. It is also noteworthy that the CoP_3 tetrahedron remains almost unaffected on bonding to the two $Cr(CO)_5$ groups.

The cyclo-P_3 unit can also behave as a three electron bridging ligand (45) to give a variety of homo and heterometallic triple-decker sandwich complexes which are treated in a recent review (100).

The remarkable ideally cubic mixed metal/pnicogen $Sb_4[Co(CO)_3]_4$ (50), or $P_4[SnPh]_4$ (51), and tetragonally distorted metal/phosphorus $P_4[CpCo]_4$ (52) cubane-like clusters (Figure 4), obtained by Dahl or Schumann, may be viewed along similar lines. In the latter case each 14-electron CpCo group completes its valence-shell by sharing electron pairs with one other CpCO group and with three phosphorus atoms, consistent with its diamagnetic character and with the observed geometry. It is likely that many more such mixed clusters, incorporating both transition metals and main group element, will be developed in the future.

A range of polyphosphorus compounds, many of which exhibit remarkable polycyclic skeletons, including stable compounds that can be formally derived from the P_4S_3 or β-P_4S_5 frames by replacing the sulfur by P^- or PR bridges, have been developped in recent years, and are treated in two fascinating recent review articles (19,20). Their coordination chemistry, which does not seem to have been investigated yet, should prove promising, especially in view of the fact that several transition metal adducts of monocyclic polyphosphines have already been reported (53).

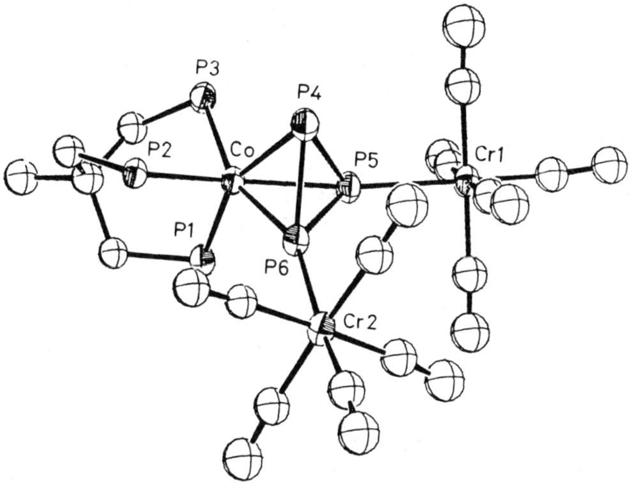

Figure 3. ORTEP drawing of $P_3(Co)(triphos)[Cr(CO)_5]_2$ (49).

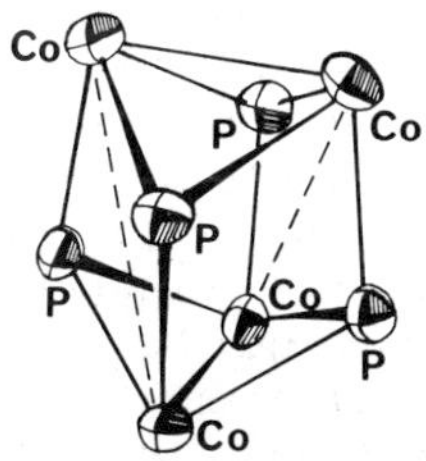

Figure 4. The $P_4(Co)_4$ core in $P_4(\eta^5 - C_5H_5Co)_4$; Cp groups are omitted (52).

Tetraphosphorus Oxides and Imides as Donors. The Transmission of Electronic and Structural Effects through the Molecular Frame

The Coordination Chemistry of P_4O_6, P_4O_7 and $P_4(NMe)_6$. The phosphorus oxides, imides, their derivatives and their analogs usually have the adamantane-type structure shown in scheme 1. All bear numerous lone pairs on their phosphorus or arsenic and/or oxygen, nitrogen, sulfur, etc.., atoms, whether in branching, bridging or terminal positions, thus offering a wide choice of potential donor sites for coordination. π-accepting character can also be anticipated from the elements from the Third Row on. No chelating behavior can be expected, however, from the branching P(III) or As(III) atoms, in view of the diverging orientations of their lone electron pairs. The participation of 3d orbitals in bonding in P_4O_6 has been judged, from an LCAO-SCF-CNDO calculation, to be substantial (54); it may provide a mechanism to even out the relative charges on the P and O atoms.

These expectations were borne out for example by the reactions of diborane and of a series of metal carbonyls with P_4O_6 and $P_4(NMe)_6$, whose coordinating abilities have been thoroughly investigated (28,55). P_4O_6, in contrast to its usual behavior, which is often characterized by violent reactions in which the polyphosphorus structure is shattered (56), reacts smoothly at room temperature, with a large excess of $Ni(CO)_4$, to add 1 to 4 $Ni(CO)_3$ groups in a sequential manner, as shown by ^{31}P NMR monitoring (55):

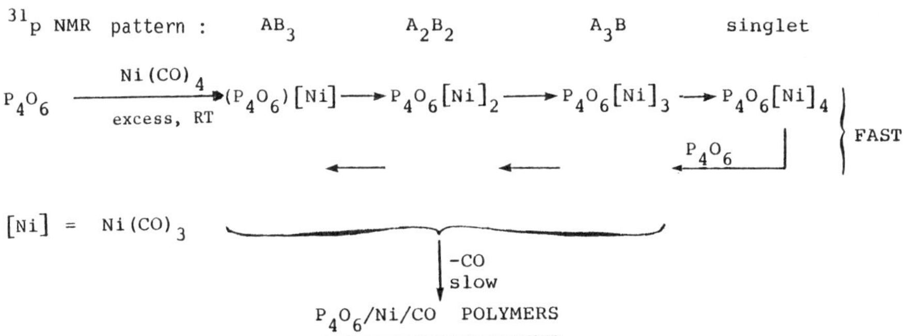

When on the contrary it is P_4O_6 that is in excess, it can replace up to three carbonyl groups in $Ni(CO)_4$. When neither reagent is in excess, highly cross-linked, three-dimensional polymeric networks are obtained, in which the nickel atoms are bridged by intact P_4O_6 molecules, which in turn are bridged by nickel atoms (55). As expected from the four-fold branching character of both P_4O_6 and the Ni atoms, there are two gel points in the system, at Ni/P_4O_6 molecular ratios of ca 0.25 and 4. It is noteworthy that polycondensation does not occur immediately when the isolated $(P_4O_6)[Ni(CO)_3]_4$ is allowed to react with P_4O_6 : the initial steps of the reaction consist of the progressive stripping, in a matter of minutes, of $Ni(CO)_3$ groups from the P_4O_6 core until only one is left, thus avoiding, at the first stage, the region of polymeric structures in which the reaction rates decrease rapidly as cross-linking increases.

Tetraphosphorus and tetraarsenic hexamethylimides similarly add one to four $Ni(CO)_3$ groups, but the coordination of more than one $P_4(NMe)_6$ ligand on Ni, and hence the formation of polymeric structures, are not observed, presumably for steric reasons (57, 58). The $Ni(CO)_3$ groups always exchange between the phosphorus donor sites; in the case of the arsenic analog $As_4(NMe)_6$, this exchange is fast enough to provoke the coalescence of the signals in the ^1H NMR (Figure 5).

The reactions of P_4O_6 (and P_4O_7) with iron and group VI carbonyls have also been investigated, principally by Mills et al. (59-61):

$$P_4O_6 + Fe(CO)_5 \xrightarrow[\text{diglyme}]{140^\circ} \left\{ \begin{array}{l} \underline{(CO)_4Fe(P_4O_6)} \\ \underline{(CO)_3Fe(P_4O_6)_2} \\ \underline{P_4O_7} \xleftarrow[\text{diglyme}]{\Delta} P_4O_6 \end{array} \right. \quad (59)$$

$$P_4O_6 + Fe(CO)_5 \xrightarrow[\text{thf, RT}]{h\nu} \underline{(CO)_4Fe(P_4O_6)} + (CO)_3Fe(P_4O_6)_2 + P_4O_9\ (?)$$

$$P_4O_6 + Fe_2(CO)_9 \xrightarrow[\text{RT, thf}]{} [(CO)_4Fe]_n(P_4O_6) \qquad n = \underline{1}, 2, 3, 4 \quad (59)$$

$$P_4O_6 + M(CO)_6 \xrightarrow[\text{or cyclohexane reflux}]{\text{thf, h}\nu} [(CO)_5M](P_4O_6) \qquad M = Cr, Mo, W \quad (60)$$

$$P_4O_7 + nM(CO)_m \qquad |M(CO)_{m-1}|_n\ (P_4O_7) + n\ CO \quad (61)$$

$M = Ni$, $n = 1$-3; $M = Fe$, $n = 1,2$; $M = Cr, Mo, W$, $n=1$

isolated

The thermal reaction of P_4O_6 with $Fe(CO)_5$ leads to a mixture of the mono and di-iron carbonyls, which were separated by sublimation; the accompanying autoxidation of P_4O_6 to P_4O_7 is presumably independent of the presence of the iron compound, since it could be obtained in high yields by simply heating a diglyme solution of P_4O_6 (59). The same metal adducts were obtained at lower temperature under photochemical activation; they were then accompanied by another uncoordinated phosphorus species, probably P_4O_9. Similar behavior was found with $Fe_2(CO)_9$, and appears to take place with $Fe_3(CO)_{12}$. The $P_4O_6[M(CO)_5]$ compounds (M = Cr, Mo, W) (60) and their P_4O_7 analogs (61) display high carbonyl stretching frequencies, consistent with the ligand's having a strong π-accepting character. The increase in frequency noted when descending the

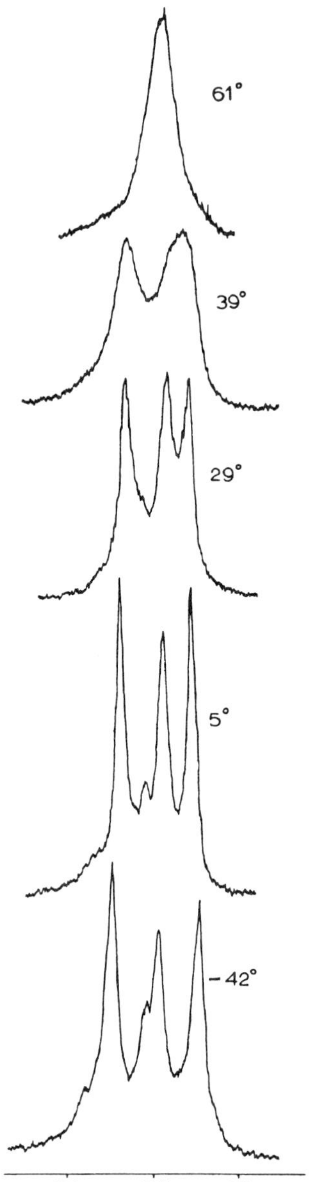

Figure 5. Variable-temperature ^1H-NMR of a mixture of $As_4(NMe)_6[Ni(CO)_3]_n$ complexes in $CDCl_3$.

VIth group is an indication of a decrease in the π^* carbonyl orbital population, as expected from the decrease of the P(3d) orbital overlap with the larger, more diffuse, 4d or 5d orbitals of Mo or W. The high value (486 Hz) of the $J(^{31}P-^{183}W)$ coupling constant, which is also thought to reflect the π-acceptor ability of phosphoranes, ranks P_4O_6 among the strongly π-accepting ligands.

Nöth and Thorn reported the preparation of $P_4(NH)_6[M(CO)_5]_4$ with M = Cr and Mo, by thermal decomposition of $P(NH_2)_3M(CO)_5$, and of $P_4(NH)_6S_4$, through the reaction of S_8 on the chromium adduct; the mixed $P_4(NH)_6[Cr(CO)_5]_nS_{4-n}$ were identified during the process (62). Since the parent compound $P_4(NH)_6$ itself is still unknown, this exemplifies the possibility of obtaining new closo-structures, or less stable isomers of known ones, in a complexed form.

The Non-independent Behavior of the four Phosphorus Atoms. The wealth of data now available on molecules having the P_4O_6 or $P_4(NMe)_6$ core provides a basis for discussing the transmission of electronic and/or steric effects through the closo-structure, which manifests itself in many ways -for example, in the comportment of P_4O_6 and $P_4(NMe)_6$ towards B_2H_6 and BF_3. Both ligands react readily with diborane under one atmosphere to give $P_4E_6(BH_3)_n$ adducts, with n = 1 to 3 for E = O, and 1 to 4 for E = NMe; coordination occurred only on phosphorus; crystalline $P_4O_6(BH_3)_2$, $P_4O_6(BH_3)_3$ and $P_4(NMe)_6(BH_3)_4$ were isolated (63-66). An interesting point, as shown in Figure 6 for $P_4(NMe)_6$, is that the composition of the mixture of adducts formed was initially close to that which would be expected from random occupation of the donor sites, but that this composition evolved slowly, as the BH_3 groups redistributed themselves among the available sites, to finally reach an equilibrium distribution that departed considerably from randomness. Such redistribution data reflect the relative thermodynamic stabilities of the various species in presence (67), as well as differences in energies of chemically equivalent bonds (68). The redistribution is faster among the $P_4C_6(BH_3)_n$ adducts, and the equilibrium distribution (Figure 7) departs even further from randomness, showing again that each phosphorus site "senses" whether the other sites are coordinated or not.

The effect of the oxidation of one phosphorus atom on the Lewis basicity of the remaining three, has been approached through the investigation of the coordination chemistry of P_4O_7 (61), and proved remarkable. Thus, in contrast to P_4O_6, tensiometric titration of P_4O_7 with B_2H_6 at ambient temperature and normal pressure indicated no reaction. On the other hand, P_4O_7 has been reported to add two equivalents of BF_3 (61), while P_4O_6 in comparable conditions is decomposed by $BF_3.OEt_2$ to a mixture which includes PF_3, PF_2OPF_2 -whose preparation by another, selective, route was reported by Ralph Rudolph (69) -, PF_2OEt and polymeric materials in which F_2BO- and FBO_2- fragments were identified (64,70). The phosphorus lone pairs on P_4O_7 should indeed be less polarizable than those of P_4O_6, due to the electron-withdrawing effect of the

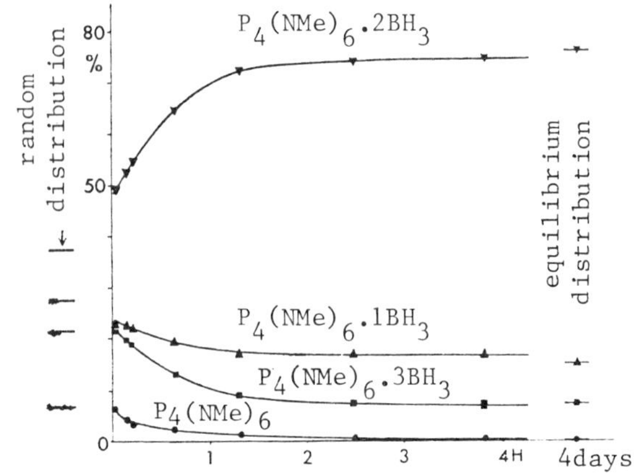

Figure 6. Evolution of the composition of a mixture of $P_4(NMe)_6(BH_3)_n$ complexes upon standing (66).

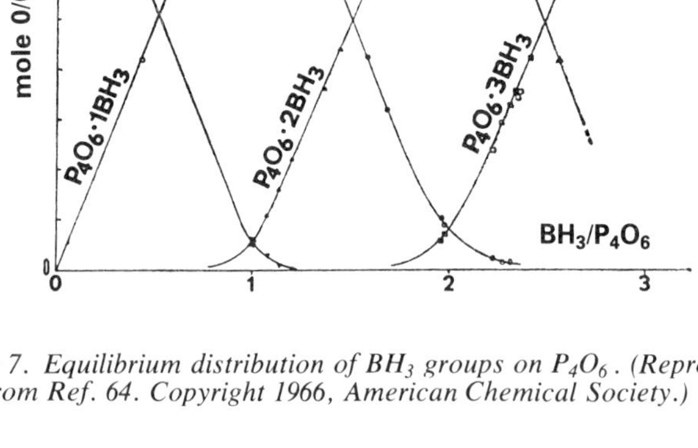

Figure 7. Equilibrium distribution of BH_3 groups on P_4O_6. (Reproduced from Ref. 64. Copyright 1966, American Chemical Society.)

exocyclic oxygen. Pertinent to the argument is also the observation that $P_4O_6S_2$ is more easily oxidized than P_4O_8 by P_4S_{10} (13). The lower reactivity observed for the residual P(III) site in $P_4(NMe)_6S_3$, compared to $P_4(NMe)_6$, under the action of MeI, PhN_3, B_2H_6 or $Ni(CO)_4$ (71), can also be examined along these lines. It can be rationalized both in terms of the electron-withdrawing effect caused by the pentavalent P's, and in addition, by the increased crowding of the P(III) site, resulting from non-bonding repulsions between the sulfur atoms and the methyl groups, with no means of apportioning these two effects.

The analysis of the changes observed in ^{31}P chemical shifts upon stepwise coordination in terms of transmission of electronic effects is a tempting approach. However, this should only be undertaken with extreme caution, particularly in view of the pronounced dependence of these shifts on angular changes (65), and of the fact that *closo*-structures are highly deformable (*vide infra*). Figure 8 shows that the chemical shifts of both tri and tetracoordinated phosphorus atoms are indeed affected by stepwise coordination, but in a rather peculiar way, with the effect often reaching a maximum after one or two substitutions, and diminishing again thereafter.

First it should be clear that the amplitude of these variations can probably be accounted for by angular changes of only a few degrees (65). Then, their unusual appearance should probably be considered together with the change in symmetry from T_d to C_{3v}, then C_{2v}, and back to C_{3v} and T_d, during the coordination sequence. The chemical shift variations may therefore primarily reflect structural deformations and angular variations, and will probably prove to be of little help in appreciating displacements of electron density during the coordination process.

One of the most obviously desirable approaches to the quantitative appreciation of such transmission effects, as well as of the extent of multiple character in bonding, would have been to perform X-ray diffraction analysis on a complete series of adducts with 0 to 4 coordinated phosphorus atoms. This has so far been prevented by difficulties in isolating the partially coordinated species, largely on account of their instability with respect to redistribution of the Lewis acids among the donor sites. But, since the problem does not change in nature when oxygen or sulfur are stepwisely added in the exocyclic positions of the *closo*-structures, instead of the previously mentioned Lewis acids, a series of oxyimides, thioimides and mixed oxythioimides was prepared for this purpose, according to following scheme.

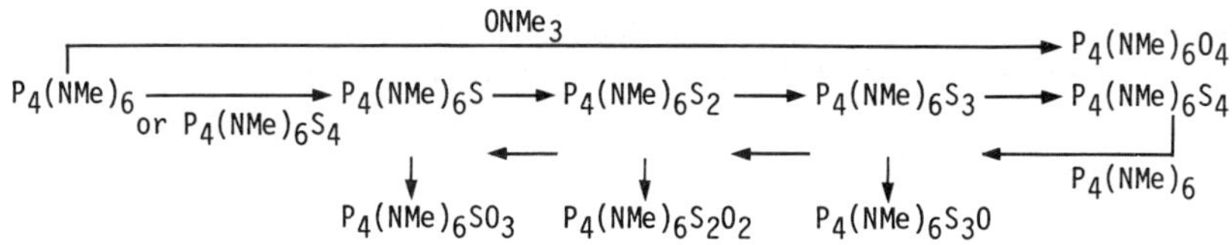

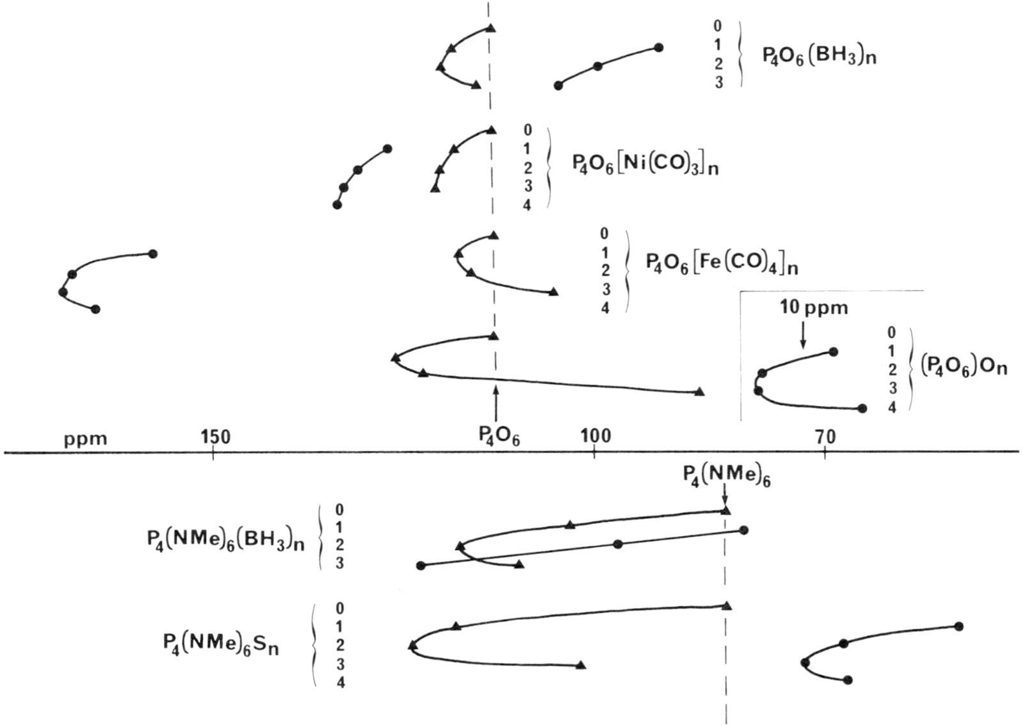

Figure 8. Variation of ^{31}*P-NMR shifts in* $P_4E_6(X)_n$ *as a function of n. Key: ●, coordinated; and ▲, uncoordinated.*

All four sulfides formed readily, were isolated, and gave single crystals (72); the tetrasulfide had been prepared previously by Holmes (73). On the contrary, only the tetraoxide, along with the complementary amount of unreacted starting material, was obtained, independently of the stoichiometry of the smooth oxidizing agent employed, $ONMe_3$; no partially oxidized imides were formed either upon attempted redistribution of the peripheral oxygens between $P_4(NMe)_6$ and $P_4(NMe)_6O_4$ (14). This is certainly another striking manifestation of the transmission of electronic effects through the structure : the coordination of a first oxygen atom would increase the affinity for oxygen of the remaining ones to such an extent that the fully oxidized species would form.

The exchange of the peripheral, formally double-bonded sulfur atoms, when for example $P_4(NMe)_6$ is added to $P_4(NMe)_6S_4$, was found to occur without loss of the inner $P_4(NMe)_6$ core (74,75). However, it takes several days at 80°C to reach an equilibrium among the $P_4(NMe)_6S_4$ molecules; the molecular distribution then again departs significantly from randomness. The fact that the term with $n = 2$ is less stable than those with $n = 1$ or 3 could also be related to changes in symmetry rather than to electronic polarization effects which would be expected to vary in a monotonous way as n increases. The same holds for P_4S_8 - or $(P_4S_6)S_2$ - which is less stable than both P_4S_7 and P_4S_9 (8).

Structural Disorder : a Prominent Feature of Closo-Compounds

X-ray structure determination have been performed on the entire series of compounds $P_4(NMe)_6S_n$ ($n = 0 - 4$) and $P_4(NMe)_6O_4$ with the aim of gaining some insight into the extent of π contribution to their bonding and the transmission of effects through structures (76-80).

Accurate bond distances were expected, if only because each type of P-N bond is repeated, for example 12 times in molecules of T_d symmetry - or even 24 or 48 times when two ($P_4(NMe)_6$) or four ($P_4(NMe)_6O_4$) crystallographically independent molecules are present in the unit cell - thus providing 12, 24 or 48 independent measurements of the same molecular quantity, from which an accurate average would ordinarily be expected.

In spite of solicitous efforts - data being in some cases collected on several crystals or at different temperatures (see also ref. 81) - only disappointingly imprecise bond distances were obtained, just sufficient to discern qualitative trends in bond-length variations upon progressive coordination of the closo structure, but insufficient to permit any valid discussion of these data in terms of the various contributions that might determine them. This tendency to disorder is illustrated by a large spread in bond length - usually over a tenth of an angström - so that the average value carries a wide interval of uncertainty (77). It also manifests itself by the existence of polymorphs, monoclinic and orthorhombic for $P_4(NMe)_6S$ (78), and, in the case of $P_4(NMe)_6S_4$, of a gross disorder (Figure 9) of a puzzling character, with at 3°C, two possible locations for the sulfur atoms with occupancy

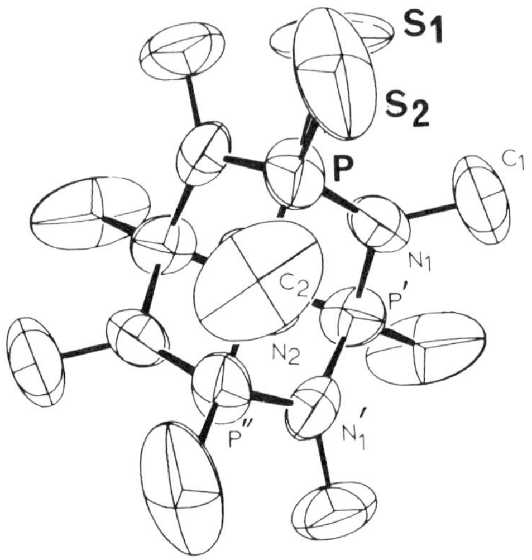

Figure 9. A view of the $P_4(NMe)_6S_4$ molecule with the two locations S_1 and S_2 found for the sulfur atoms at 3 °C (77). This disorder is shown on one phosphorus atom only to avoid overcrowding.

factors close to 50% (three likely positions for these sulfurs, with occupancy factors of 0.65, 0.28 and 0.07%, were found at 22°C (77).

The most interesting outcome of this study lies in fact in the recognition that this uncertainly and tendency to structural disorder appears to be a common, prominent feature in closo-phosphorus compounds.

This feature can be understood intuitively on two grounds :
1) the closo-molecules have no central atom tying the atoms together across the cage structure, and hence should be envisioned as rather soft, deformable envelopes (although there is no room inside to accommodate such a central atom), reminiscent of a deflated tennis ball, subject to easy deformation under the action of crystal packing forces; as a result the electron distribution is more likely to adapt to, rather than to determine, the geometric structures.
2) these molecules are close-to-spherical in shape - or globular - which tends to favor rotational disorder and/or librational motion of large amplitude in the crystal, and, in the extreme cases, leads to the existence of a distinct plastic phase characterized by virtually free rotation in the solid. Therefore only qualitative trends can be recognized and discussed (79-80).

Trends in P-O and P-N Bond Lengths; Contribution to π-Bonding

The bond length data collected for $(P_4O_6)_n$ and for $P_4(NMe)_6S_n$ are displayed diagrammatically below (Å) :

P_4O_6 / $P_4(NMe)_6$:
$$P(III){-}E \frac{1.638(3)(gas)\;\; 1.656(4)(solid)}{1.70(1)} P(III)$$

P_4O_7 / $P_4(NMe)_6S$:
$$P(III){-}E \frac{1.645(7)}{1.70(1)} P(III) \frac{1.684(7)}{1.73(2)} E \frac{1.593(7)}{1.63(2)} P(V)$$

P_4O_8 / $P_4(NMe)_6S_2$:
$$P(III){-}E \frac{1.63(1)}{1.70(1)} P(III) \frac{1.67(1)}{1.73(1)} E \frac{1.57(2)}{1.66(1)} P(V) \frac{1.59(1)}{1.692(4)} E{-}P(V)$$

P_4O_9 / $P_4(NMe)_6S_3$:
$$P(III) \frac{1.66(1)}{1.71(2)} E \frac{1.60(2)}{1.65(2)} P(V) \frac{1.59(1)}{1.68(2)} E{-}P(V)$$

P_4O_{10} / $P_4(NMe)_6S_4$:
$$P(V) \frac{1.60(1)}{1.66(1)} E{-}P(V)$$

(Data from ref. 4, 76-79, 82-85).

In view of the large esd's usually associated with these data, differences among individual bond lengths may not always be significant. But the fact that parallel trends emerge throughout both series substantiates their reality.

The shortening of the P-E bonds when P becomes pentavalent is surprisingly low and often not significant. Such shortening was certainly expected on either or both of two grounds : first, the increase in formal oxidation state of the phosphorus atom upon attachment of an exocyclic oxygen or sulfur atom should lead to a decrease in its covalent radii; data available on acyclic analogs suggest that this mechanism might account for 0.03-0.04 Å of shortening (79). Then the increase in π-electron density drawn from the nitrogen atoms by the more positive phosphorus atom could also contribute in strengthening the bond; the contraction of the d orbitals should in particular enhance the Np_{π}-Pd_{π} contribution to bonding.

More interesting, and revealing of a second range effect, is that the nature of the phosphorus atom located on the other side of the bridging atom has a clear effect on the P-N- or P-O- bond length. The general trend is the following (see in particular P_4O_8 and $P_4(NMe)_6S_2$ which contain all four possible arrangements) :

$$P(III)\text{-}EP(V) > P(III)\text{-}EP(III) > P(V)\text{-}EP(V) > P(V)\text{-}EP(III)$$

The longest and the shortest bonds are both found in the P(III)-E-P(V) sequence. The first gains ca 0.03 Å, while the second loses a comparable length with respect to the "same" bond as found in the symmetric P_4E_6 and $(P_4E_6)E'_4$ molecules. The electrons are attracted on the side of the pentavalent phosphorus at the expense of the bond located on the other side of the bridging element; this displacement of electron density can occur through both the σ and the π bonding frames.

X-Ray studies on the phosphorus oxides led to similar observations (82-86). P_4O_7, for example, cannot be considered as made up of fragments taken from P_4O_6 and P_4O_{10} (85). They also reveal the effect of packing, as well as significant differences between the solid and gas phase structures (84,86).

Evidence for Np_{π}-Pd_{π} contribution to bonding in $P_4(NMe)_6S_n$ include the close-to-planar arrangement of the bonds about all nitrogen atoms (deviations are in the direction expected from steric effects rather than related to the type of bonded P atoms), the significant shortening - by 0.07-0.10 Å - of the P-N bonds, with respect to properly redefined $P(III)\text{-}N(sp^2)$ and $P(V)\text{-}N(sp^2)$ standards (taking into account both the change in hybridization of N with respect to the usual standards, and the oxidation state of phosphorus (79), the high energy and breadth of the nitrogen atom's lone pair absorptions in the photoelectron spectra, and, to a lesser degree, some NMR coupling data.

With an average 1.70 Å, all P-N bonds are shorter than would be expected on the basis of σ bonding alone, even after the usual single bond radii are corrected for the change of the nitrogen's hybridization from sp^3 to sp^2. This shortening would correspond, on the basis of Cruickshank's treatment (83), to a bond-order in the range of 1.18 to 1.30 for the P(III)—N-P(III) bonds.

The π contribution to bonding may in fact be larger than would appear from the bond shortening. The fact that there is no significant change in P-N bond-lengths in the series $P_4(NMe)_6$, $P_4(NMe)_6S_4$ and $P_4(NMe)_6O_4$ may signify that part of the bond-length's shortening, which would normally result from a given π contribution in an open system, is absorbed here in counteracting increased Van der Walls repulsion between non-bonded atoms in the closed, constrained, polycyclic system.

The absorption of the nitrogen's lone-pairs of $P_4(NMe)_6$ in the photoelectron spectrum lies between 8 and 9 eV, while it would have been expected to be well below 7 eV for a pure N2p lone pair. And the breadth of the bond indicates that these electrons are not simply isolated, equivalent lone pairs, but that they occupy delocalized orbitals with bonding character (76). The addition of 1 to 4 exocyclic sulfur atoms provokes a regular increase in the ionisation energy of these Npπ electrons, the overall shift being about 1 eV from $P_4(NMe)_6$ to $P_4(NMe)_6S_4$. This stabilization of the electrons is consistent with a strengthening of the Npπ-Pdπ interaction, but could also arise indirectly by polarizing the N-P bonds so as to cause the more positive nitrogen atoms to bind the P electrons more strongly.

In the NMR spectra, the ^{13}C-^{1}H coupling constants increase from $P_4(NMe)_6$ to $P_4(NMe)_6S_4$ to $P_4(NMe)_6O_4$; this is consistent with the expectation that the exocyclic atoms will draw electron density away from the nitrogen atoms, and this to a greater extent with oxygen than with sulfur, and also consistent with increased Np_π-Pd_π bonding, in the same order (80). But this interpretation again is not unique, since the attraction of electron density can occur either through the σ system or through the π system, or both, and the NMR data provide no basis for determining the relative importance of the two pathways.

Plastic phases. Plastic phase behavior has been recognized for P_4, P_4S_3, P_7^{3-}salts, P_4S_{10}, $P_4(NMe)_6$ and $As_4(NMe)_6$. It is exemplified here by $P_4(NMe)_6$ (87). Figure 10 shows the variation in line width of the signal in the ^{1}H NMR spectra measured on solid $P_4(NMe)_6$ between 40 and 140°C. The fact that the line width is "only" 4700 Hz at 40°C already supposes considerable rapid reorientational movement in the solid at that temperature. The abrupt change that occurs within one degree at 56°C reflects a drastic change in molecular freedom, which was assigned to a plastic phase transition above which the molecules rotate almost freely while maintaining their positional order in the solid. This interpretation is supported by differential scanning calorimetry (Figure 11), which shows, at the same temperature, a sharp, first-order transition upon heating (upon cooling, the transition is found at 50.5°C, which denotes hysterysis, a common feature of plastic phase transitions (87)). A transition at 38°C was tentatively assigned to the rotation of the molecule about one of its C_3 axes; the fact that the process is high- order upon heating means that the onset of the rotation is concerted. The low entropy of melting

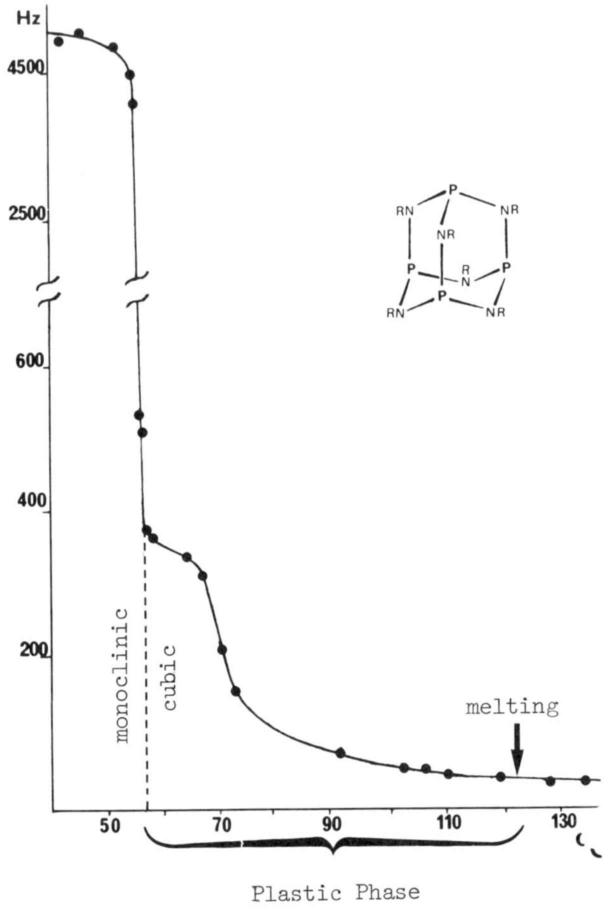

Figure 10. Line-width of the ^{1}H-NMR signal measured on solid $P_4(NMe)_6$ as a function of temperature.

(Reproduced with permission from Ref. 88. Copyright 1977, Institut Mondial du Phosphate.)

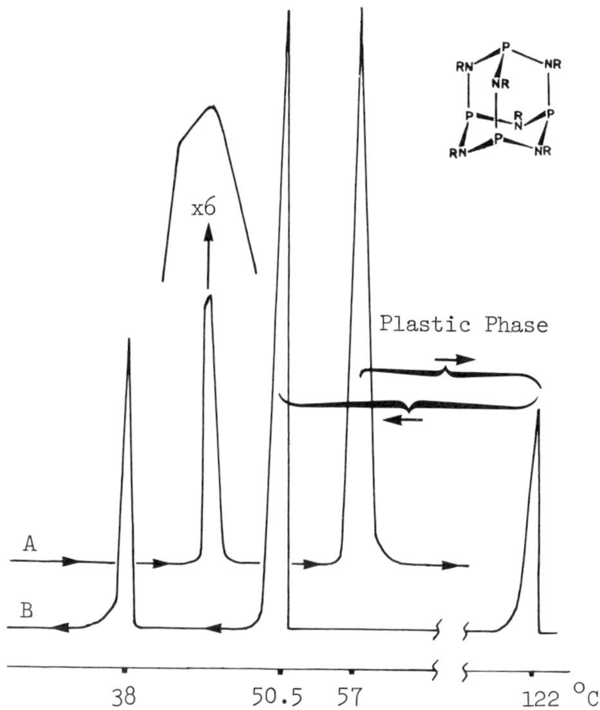

Figure 11. Differential scanning calorimetry of $P_4(NMe)_6$.
(Reproduced with permission from Ref. 88. Copyright 1977, Institut Mondial du Phosphate.)

(ΔS = 4.8 J $mol^{-1}K^{-1}$), at 122°C, is also characteristic of a plastic crystalline phase, as is the softness of the crystals and their sensitivity to macroscopic mechanical damage. Variable temperature X-ray powder diffraction shows a change in pattern from monoclinic to cubic upon heating; the diffraction lines, sharp and numerous for the "rigid" crystals, become few and less well-defined for the plastic crystal.

It should be noted that the structural criteria - globularity - which is usually sufficient to determine the existence of a plastic phase for organic compounds, no longer suffices here, since P_4, P_4S_{10} and $P_4(NMe)_6$ do exhibit plastic phases, while the equally globular P_4O_6 and P_4S_{10} of same symmetry do not (89). Several factors are likely to contribute to this difference. First the intermolecular hydrogen/hydrogen interactions,which are known to have a determining influence on the crystal packing, should contribute to the separation of molecules such as $P_4(NMe)_6$, and thus favor free rotation in the solid; $P_4(NMe)_6$ is almost perfectly spherical, and literally coated with hydrogen atoms, as is adamantane. On the other hand, the location and local polarity of the oxygen atoms in P_4O_6 and P_4O_{10} probably contribute to increasing molecular interlocking, and therefore hinder the rotation. Thus the way the P_4O_{10} molecules pack together is of much interest: each molecule has four spikes and four recesses, and the spikes of each molecule fit into the recesses of its four nearest neighbors, leading to a close packing with considerable interlocking. The structural data show that such interpenetration does not exist in P_4S_{10}, with its peripheral sulfur atoms too large to fit into the nitches of the neighboring molecules, and with smaller dipolar interactions. These observations led to the defining of a new structural criterion for predicting plastic phase behavior, based on the degree of intermolecular interlocking (89).

The Coordination Chemistry of Tetraphosphorus Sulfides : a Deceptive but Open Case

The *closo*-tetraphosphorus sulfides, with their various stoichiometries and unique structures (Scheme 2), exhibit multiple and often distinct potential coordination sites. These sites often involve several distinct sets of formally non-bonding electron pairs at both phosphorus and sulfur atoms, and the tetraphosphorus sulfides could therefore be expected to exhibit a wide range of coordination modes and original behavior towards Lewis acids. But so far this has not proved so.

This is all the more disappointing, as a thorough theoretical investigation of the electronic structure of P_4S_3, performed by Xα scattered wave, as well as by the extended Hückel and CNDO molecular orbital methods, led to the conclusion that the apical and the basal phosphorus atoms could be expected to be equally good donors (90). The electronic structure of the molecule is described by a set of lower energy valence-shell levels having high 3s character, which provide its main cohesive energy through multi-center bonding within the P_3 and PS_3 sub-units, and by another set of higher, occupied non-bonding or weakly bonding levels that have a large 3p component. Contour plots show the latter to have electron distributions characteristic of directed lone pairs both at the apical P and basal P_3- unit. While the apical phosphorus is predicted to have donor properties similar to those of tertiary phosphanes, the basal P_3 unit would be expected to behave essentially like P_4.

The high ionization energy (from HeI photoelectron spectroscopy (90)) assigned to the apical lone pair electrons indicates a very weak base. This is consistent with the lack of reactivity of P_4S_3 towards BCl_3 or BF_3, its failure to be quaternized under the action of $MeSO_3F$ or Et_3OBF_4 and, in general, its resistance to protonation (91).

A first series of transition metal adducts of P_4S_3 was obtained in 1969 by Nixon et al. from the following reactions (92) :

$$2\ P_4S_3 + C_7H_8M(CO)_4 \xrightarrow{CS_2\ \text{reflux}} cis\text{-}(P_4S_3)_2M(CO)_4 \quad M = Cr, Mo, W$$

$$3\ P_4S_3 + C_7H_8M(CO)_3 \longrightarrow fac\text{-}(P_4S_3)_3M(CO)_3 \quad M = Cr, Mo$$

$$4\ P_4S_3 + (\eta^5\text{-}C_5H_5)_2Ni \longrightarrow (P_4S_3)_4Ni$$

$$4\ P_4S_3 + (\eta^3\text{-}C_3H_5)_2Ni \xrightarrow{-30°,\ Et_2O} (P_4S_3)_4Ni$$

cis and *fac* stereochemistries were proposed on the basis of the IR spectra. The ^{31}P spectra were reported as consisting of a doublet and a quartet in 3:1 ratio, indicating that (on the average) the C_{3v} symmetry of the ligand had been retained. Both signals were shifted towards lower fields with respect to those observed for free P_4S_3. As the largest shift affected the quartet, it was proposed that it is the apical phosphorus which is bound to the metal. $Ni(P_4S_3)_4$ was obtained by displacement of both cyclopentadienyl rings from nickelocene or, more conveniently, from bis-π-allyl nickel (92).

$P_4S_3Mo(CO)_5$ was prepared by Cordes et al. by refluxing equimolar amounts of P_4S_3 and $Mo(CO)_6$ in cyclohexane containing a small amount of diglyme (93). The yield was improved by photochemical activation; $P_4S_3W(CO)_5$ was also thus obtained in 66% yield (94). The X-ray crystal structure of the former reveals that, in the solid, the intact P_4S_3 molecule behaves as a monodentate ligand through its apical phosphorus atom (Figure 12) (93). The P_4S_3 core shows surprisingly little deformation upon coordination : none of the bond angles nor the basal bond lengths are changed at all, and only barely significant lengthening of both types of P-S bonds could be noted. The short Mo-P distance of 2.477(9) Å, the high carbonyl stretching frequencies and an Mo-C bond length comparable to those in $Mo(CO)_6$ and $Mo(CO)PF_3$, lead to the ranking of P_4S_3 among the strongly π-accepting ligands.

Insoluble, probably polymeric adducts, formulated as P_4S_3CuX (X = Cl, Br, I), have been obtained by allowing P_4S_3 to react with Cu(I) or Cu(II) halides in CS_2 (95).

Coordination through the basal phosphorus atoms has so far not been established, in spite of numerous attempts. The ^{31}P NMR spectra displayed by $P_4S_3M(CO)_5$ is however puzzling (94). These spectra independent of solvent (hexane, toluene, CH_2Cl_2, CS_2), and metal (Mo, W), exhibit *two* sets of signals having *ca* 3:1 peak area ratio, each consisting of a doublet and a quadruplet, also in *ca* 3:1 ratio. The most intense of these sets has chemical shifts and coupling constants identical to those of free P_4S_3(δ = -128.4(d), +62.1(q) ppm, J = 70 Hz). The second, weaker, set is found at lower field, and shows a greatly reduced splitting (Mo : δ = -111.9(d), +122.5(q) ; J = 33 Hz; W : δ = -111.3(d), +84.8(q) ppm; J = 30 Hz). Both the solution and solid state infra-red spectra show no vibrations in the carbonyl region other than those characteristic of an $M(CO)_5$ moiety. When the solvents were evaporated, the initial crystalline adduct could be recovered.
Possible coordination by the sulfur atoms seems to be excluded by the Mo-P coupling constant observed in the ^{95}Mo NMR (J_{Mo-P} = 140Hz at δ = +99 ppm) with respect to $Mo(CO)_6$). The addition of free PPh_3 had no influence on the process and PPh_3 remains uncoordinated in a solution of $P_4S_3Mo(CO)_5$. A dissociation to form an apex-plus-base coordinated $P_4S_3[M(CO)_5]_2$ adduct which would retain the C_{3v} symmetry of P_4S_3 is hardly compatible with the peak area ratio found in the NMR. Thus, the hypothesis we favor, on the basis of the peak area ratios, would be a dissociation according to :

$$4\ P_4S_3M(CO)_5 \longrightarrow P_4S_3[M(CO)_5]_4 + 3\ P_4S_3$$

in which p_4S_3 would behave as a *tetra*dentate, apex-plus-base coordinated ligand; but this remains to be established.

The ligand behavior of the arsenic-based compound **9** (Figure13) structurally related to P_4S_3, is of interest in this context. It was shown that **9** reacts with group VI metal hexacarbonyls under UV irradiation to give the adducts $[CH_3C(CH_2As)_3]_nM(CO)_{6-n}$ (N = 1, M = Cr, Mo, W; n = 2, M = Cr, W) in which **9** behaves as a monodentate ligand (96). An X-ray diffraction analysis (n = 1, M = Cr) definitely proves that in the solid the metal is located on one of the three arsenic atoms of the base, reminiscent of P_4 adducts (97). Polymeric species of formula $[CH_3C(CH_2As)_3Mo(CO)_4]_n$ and $[CH_3C(CH_2As)_3Mo(CO)_3]_n$, for which bridging bidentate behavior of the ligand was suggested, were also obtained, and a cobalt carbonyl nitrosyl adduct, $CH_3C(CH_2As)_3CO(NO)(CO)_2$, was briefly mentionned (96).

Related to the coordination chemistry of P_4S_3 is that of α-$P_4S_3I_2$, which also displays multiple potential coordination sites in a semi-open (nido) structure, but with fewer steric constraints and no P_3-unit. The coordination abilities of α-$P_4S_3I_2$, however, have also proven rather limited. Baudler *et al* reported a series of molecular low-valent metal adducts of type $M(\alpha\text{-}P_4S_3I_2)_n(CO)_{n-m}$(M = Ni, Fe; m = 1-3) (98). It is remarkable - and frus-

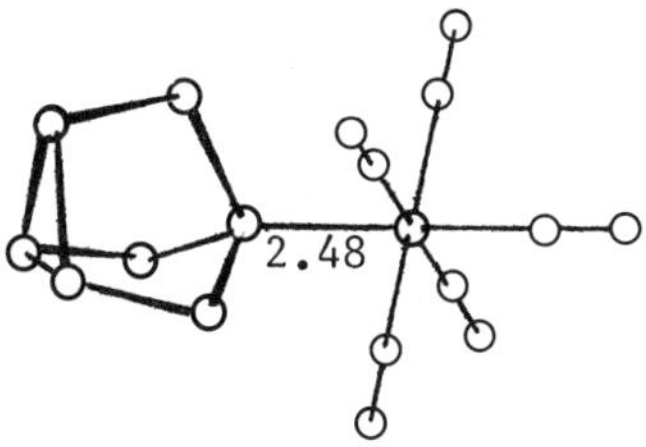

Figure 12. x-Ray structure of $P_4S_3Mo(CO)_5$ (93).

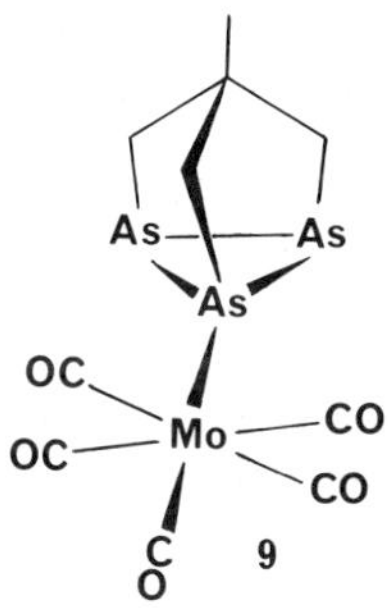

Figure 13. Structure of $CH_3C(CH_2As)_3Cr(CO)_5$ (97).

trating that the ligand acted exclusively as a monodentate ligand; whether this arises from steric hindrance or from electron delocalisation within the ligand's frame is not clear. Depending on the solvent used, $[Co(CO)_4]_2$ reacts with α-$P_4S_3I_2$ to give the substitution product $[Co(\alpha\text{-}P_4S_3I_2)(CO)_3]_2$ or an ionic complex $[Co(\alpha\text{-}P_4S_3I_2(CO)_3]^+$ $[Co(CO)_4]^-$ via valence disproportionation.

Preliminary but encouraging results have been obtained with P_4S_7 (94). When allowed to react with an excess of $Mo(CO)_6$ at reflux in cyclohexane, the insoluble P_4S_7 is converted into a yellow crystalline air-sensitive material, soluble in most standard organic solvents and with an elemental analysis which points to the formulation $P_4S_7[Mo(CO)_5]_n$, n being 6 or 7! Single crystals were obtained from CH_2Cl_2, but decomposed with release of $Mo(CO)_6$ upon standing, or under the X-ray beam. The presence of $Mo(CO)_5$ groups (and absence of $Mo(CO)_6$)in the adduct was established by the characteristic ν(CO) absorption pattern and ν(MC) and ν(MCO) vibrations, in the infra-red, and is further confirmed by the ^{17}O NMR, which exhibits two signals in 4:1 ratio. The ^{31}P NMR ($CHCl_3$) displays two sharp signals at +107.0 and +46.4 ppm with peak area in 1:3 ratio and no detectable coupling, in contrast to +110.7 and +83.9 ppm in 1:1 ratio (also with no detectable coupling), for free P_4S_7 in CS_2 solution (99). The 1:3 signal ratio of the new adduct is no longer compatible with the initial structure of P_4S_7, and could indicate a rearrangement in which one of the terminal sulfur atoms would be inserted into the already distended (2.33 Å) P-P bond of P_4S_7. Such a rearrangement has already been found to occur between α-P_4S_5 and β-P_4S_5 (Scheme 2); further work is under way to elucidate this riddle.

When P_4S_9 and P_4S_{10} were allowed to react with $Mo(CO)_6$, sulfur was released, and a mixture of the above $P_4S_7[Mo(CO)_5]_{6(or 7)}$ -formulated adduct, with a new adduct, (which in the ^{31}P NMR displays a quartet and a doublet at 80.0 and +63.4 ppm respectively, in 1:3 ratio, with J_{PP} = 6 Hz, tentatively assigned to a $P_4S_9Mo(CO)_5$ adduct) was obtained (94).

Literature Cited

1. Van Wazer, J.R. "Phosphorus and its Compounds", Vol.I, Interscience : New York, 1958.
2. Haiduc, I. "The Chemistry of Inorganic Ring Systems", Wiley : London, 1970.
3. Armitage, D.A. "Inorganic Rings and Cages"; Arnold : New-York, 1972.
4. Corbridge, D.E.C. "The Structural Chemistry of Phosphorus" : Elsevier : Amsterdam, 1974.
5. Hoffman, H.; Becke-Goehring, M. in "Topics in Phosphorus Chemistry", Vol. 8, Griffith E.J. and Grayson, M. Ed.; Interscience : New York, 1976.
6. Heal, H.G. "The Inorganic Heterocyclic Chemistry of Sulfur, Nitrogen and Phosphorus"; Academic Press : London, 1980.

7. Scherer, O.J.; Andres, K.; Krüger, C.; Tsay, Y-H.; Wolmerhaüser, G. Angew. Chem. Int. Ed., 1980, 19, 571.
8. Barieux, J-J.; Demarcq, M.C. J.Chem.Soc.Chem.Commun. 1982,176.
9. Dahl, A.R.; Norman, A.D.; Shenay, H.; Schaeffer, R. J. Am. Chem. Soc. 1975, 97, 6365.
10. Roesky, H.W.; Tebbe, F.N.; Muetterties, E.L. Inorg. Chem. 1970, 9, 831.
11. Glonek, T.; Van Wazer, J.R.; Myers, T.C. Inorg. Chem. 1975, 14, 1597.
12. Kopf, J.; Von Deuten, K.; Klar, G. Inorg.Chim.Acta 1980,38,67.
13. Walker, M.L.; Peckenpaugh, D.E.; Mills, J.L. Inorg. Chem. 1979, 18, 2792.
14. Casabianca, F.; Pinkerton, A.A.; Riess, J.G. Inorg. Chem. 1977, 16, 864.
15. Penney, G.J.; Sheldrick, G.M. J. Chem. Soc.(A) 1971, 243.
16. Blachnik, R.; Hoppe, A.; Rabe, U.; Wickel, U. Z. Naturforsch. 1981, 36b, 1493.
17. Christian, B.H.; Gillespie, R.J.; Sawyer, J.F. Inorg. Chem. 1981, 20, 3410.
18. Bock, H.; Solouki, R.; Fritz, G.; Hölderich, W. Z. Anorg. Allg. Chem. 1979, 458, 53.
19. Von Schnering, H.G. Angew. Chem. Int. Ed. 1981, 20, 33.
20. Baudler, M. Angew. Chem. Int. Ed. 1982, 21, 492.
21. Griffin, A.M.; Minshall, P.C.; Sheldrick, G.M. J. Chem. Soc. Chem. Commun. 1976, 809.
22. Blachnik, R.; Rabe, U. Z. Anorg. Allg. Chem. 1980, 461, 87.
23. Chang, C.C.; Haltiwanger, R.C.; Norman, A.D. Inorg. Chem. 1978, 17, 2056.
24. Bues, W.; Somer, M.; Brockner, W. Z. Anorg. Allg. Chem. 1981, 476, 153.
25. Griffin, A.M. Thesis, Cambridge, U.K., 1976.
26 Somer, M.S. Dissertation, Clausthal, RFA, 1979.
27. Wibbelmann, C.; Brockner, W. Z. Naturforsch. 1981, 36a, 836.
28. Riess, J.G.; Van Wazer, J.R. J. Am. Chem. Soc. 1965, 87, 5506.
29. Pierron, E.D.; Wheatley, P.J.; Riess, J.G. Acta Cryst. 1966, 21, 288.
30. Archibald, R.M.; Perkins, P.G. Chem. Commun. 1970, 569.
31. Osman, R.; Coffey, P.; Van Wazer, J.R. Inorg.Chem. 1976,15,287.
32. Ginsberg, A.P.; Lindsell, W.E. J. Am. Chem. Soc.1971,93, 2082.
33. Von Schmid, G.; Kempny, H-P. Z.Anorg.Allg.Chem. 1977,432, 160.
34. Dapporto, P.; Midollini, S.; Sacconi, L. Angew. Chem. Int. Ed. 1979, 18, 469; Dapporto, P.; Sacconi, L.; Stoppioni, P.; Zanobini, F. Inorg. Chem. 1981, 20, 3834.
35. Foust, A.S.; Foster, M.S.; Dahl, L.F. J. Am. Chem. Soc. 1969, 91, 5631.
36 Ellis, J.E. J. Organometal. Chem. 1975, 86, 1.
37. Foust, A.S.; Foster, M.S.; Dahl, L.F. J. Am. Chem. Soc. 1969,91,5633.
38. Vizi-Orosz, A.; Pályi, G.; Markó, L. J. Organometal. Chem. 1973, 60, C25.

39. Vizi-Orosz, A. J. Organometal. Chem. 1976, 111, 61.
40. Vizi-Orosz, A.; Galamb, V.; Pályi, G.; Markó, L. J. Organometal. Chem. 1976, 107, 235.
41. Vizi-Orosz, A.; Galamb, V.; Ötvös, I.; Pályi, G.; Markó, L. Transition Met. Chem. 1979, 4, 294.
42. Malisch, W.; Panster, P. Angew. Chem. Int. Ed. 1976, 15, 618.
43. Vizi-Orosz, A.; Galamb, V.; Pályi, G.; Markó, L. J. Organometal. Chem. 1981, 216, 105.
44. Seyferth, D.; Merola, J.S.; Henderson, R.S. Organometallics 1982, 1, 859.
45. Di Vaira, M.; Ghilardi, C.A.; Midollini, S.; Sacconi, L. J. Am. Chem. Soc. 1978, 100, 2550.
46. Cecconi, F.; Dapporto, P.; Midollini, S.; Sacconi, L. Inorg. Chem.1978, 17, 3292.
47. Bianchini, C.; Mealli, C.; Meli, A.; Sacconi, L. Inorg. Chim. Acta, 1979, 37, L453.
48. Huttner, G.; Müller, H-D.; Frank, A.; Lorenz, H. Angew. Chem. Int. Ed. 1975, 14, 572.
49. Ghilardi, C.A.; Midollini, S.; Orlandini, A.; Sacconi, L. Inorg. Chem. 1980, 19, 301.
50. Foust, A.S.; Dahl, L.F. J. Am. Chem. Soc. 1970, 92, 7337.
51. Von Schumann, H; Benda, H. Angew. Chem. 1968, 20, 846.
52. Simon, G.L.; Dahl, L.F. J. Am. Chem. Soc. 1973, 95, 2175.
53. West, B.O. in "Homoatomic Rings, Chains and Macromolecules of Main-Group Elements", Rheingold, A.L., Ed.; Elsevier : Amsterdam, 1977; Chapter 18.
54. McAloon, B.J.; Perkins, P.G. Theoret. Chim. Acta (Berl.) 1972, 24, 102.
55. Riess, J.G.; Van Wazer, J.R. J. Am. Chem. Soc. 1966, 88, 2166.
56. Riess, J.G.; Van Wazer, J.R. Inorg. Chem. 1966, 5, 178.
57. Riess, J.G.; Van Wazer, J.R. Bull. Soc. Chim. Fr. 1966, 1846.
58. Riess, J.G.; Van Wazer, J.R. J.Organometal.Chem. 1967,8, 347.
59. Walker, M.L.; Mills, J.L. Inorg. Chem. 1975, 14, 2438.
60. Walker, M.L.; Mills, J.L. J. Organometal. Chem. 1976,355, 355.
61. Walker, M.L.; Mills, J.L. Inorg. Chem. 1977, 16, 3033.
62. Nöth, H.; Thorn, V. Z. Naturforschung B 1981, 36, 1424.
63. Kodama, G.; Kondo, H. J. Am. Chem. Soc. 1966, 88, 2045.
64. Riess, J.G.; Van Wazer, J.R. J. Am. Chem. Soc. 1966,88, 2339.
65. Riess, J.G.; Van Wazer, J.R. J. Am. Chem. Soc. 1967, 89, 851.
66. Riess, J.G.; Van Wazer, J.R. Bull. Soc. Chim. Fr. 1968, 3087.
67. Riess, J.G. Ann. New York Acad. Sci. 1969, 159, 174.
68. Elkaïm, J-C.; Riess, J.G. Tetrahedron, 1981, 37, 3203.
69. Rudolph, R.W.; Taylor, R.C.; Parry, R.W. J. Am. Chem. Soc. 1966, 88, 3729.
70. Riess, J.G. Rev. Chim. Minér. 1969, 6, 643.
71. Casabianca, F.; Riess, J.G. unpublished.
72. Wolff, A.; Riess, J.G. Bull. Soc. Chim. Fr. 1973, 1587.
73. Holmes, R.R.; Forstner, J.A. Inorg. Chem. 1963, 2, 377.
74. Riess, J.G.; Wolff, A. J. Chem. Soc. Chem. Commun. 1972, 1050.
75. Elkaïm, J-C.; Wolff, A.; Riess, J.G. Phosphorus 1973, 2, 249.

76. Cotton, F.A.; Troup, J.M.; Casabianca, F.; Riess, J.G. Inorg. Chim Acta 1974, 11, L33.
77. Casabianca, J.; Cotton, F.A.; Riess, J.G.; Rice, C.E.; Stults, B.R. Inorg. Chem. 1978, 17, 3232.
78. Cotton,F.A.; Riess, J.G.; Rice, C.E.; Stults, B.R. Inorg. Chem. 1978, 17, 3521.
79. Cotton, F.A.; Riess, J.G.; Rice, C.E.; Stults, B.R. Inorg. Chem. 1982, 21, 3123.
80. Cotton, F.A.; Riess, J.G.; Stults, B.R. Inorg.Chem.(submitted).
81. Hunt, G.W.; Cordes, A.W. Inorg.Nucl.Chem.Letters 1974,10,637.
82. Cruickshank, D.W.J. J. Chem. Soc. 1961, 5486.
83. Beagley, B.; Cruickshank, D.W.J.; Hewitt, T.G.; Jost, K.H. Trans. Faraday Soc. 1969, 65, 1219.
84. Jansen, M.; Voss, M.; Deiseroth, H-J. Angew. Chem. Int. Ed. 1981, 20, 965.
85. Jansen, M.; Voss, M. Angew. Chem. Int. Ed. 1981, 20, 100.
86. Jansen, M.; Möbs, M. Z.Kristallographie 1982, 159, 283.
87. Postel, M.; Casabianca, F.; Riess, J.G. Inorg. Chim. Acta. 1976, 17, L23.
88. Riess, J.G.; Postel, M.; Jeanneaux, F.; Cotton, F.A.; Stultz, B.R.; Rice, C.E. Proceed. Intl. Cong. Phosphorus Compounds (Rabat, Oct. 1977), p. 51.
89. Postel, M.; Riess, J.G. J. Phys. Chem. 1977, 81, 2634.
90. Head, J.D.; Mitchell, K.A.R.; Noodleman, L.; Paddock, N.L. Can. J. Chem. 1977, 55, 669.
91. Le Geyt, M.R. Thesis, U. of British Columbia, Canada, 1974.
92. Jefferson, R.; Klein, H.F.; Nixon, J.F. J. Chem. Soc. Chem. Commun. 1969, 536.
93. Cordes, A.W.; Joyner, R.D.; Shores, R.D.; Dill, E.D. Inorg. Chem. 1974, 13, 132.
94. Bourgund, P.; Brevard, C.; Hubert-Pfalzgraf, L.G.; Riess, J.G. Unpublished results.
95. Ibáñez, W.; Gonzalez, M.; Clavijo, C. Z. Anorg. Allg. Chem. 1977, 432, 253.
96. Ellermann, J.; Lindner, H.A.; Gäbelein, H. J. Organometal. Chem. 1979, 172, 39.
97. Ellermann, J.; Lindner, H.A.; Schössner, H.; Thiele, G.; Zoubek, G. Z. Naturforsch. 1978, 33b, 1386.
98. Baudler, M.; Mozaffar-Zanganeh, H. Z. Anorg. Allg. Chem. 1976, 423, 193.
99. Brevard, C.; Demarcq, M. Chem. Phys. letters 1981, 82, 167.
100. Di Vaira, M.; Sacconi, L. Angew. Chem. Int. Ed. 1982, 21, 330.

RECEIVED April 8, 1983

3

Cyclic and High-Polymeric Phosphazenes as Carrier Molecules for Carboranyl, Metallo, or Bioactive Side Groups

H. R. ALLCOCK

The Pennsylvania State University, Department of Chemistry, University Park, PA 16802

Cyclic and high polymeric phosphazenes can be modified by nucleophilic-type substitution reactions to generate a wide range of derivatives. Recent developments include the introduction of bioactive organic residues to yield biologically-active high polymers and the synthesis of transition metal derivatives of phosphazenes. In addition, hybrid phosphazene-carborane compounds have been prepared including examples in which nido-carboranyl units, attached to a phosphazene ring or chain, function as binding sites for transition metal organometallic units.

Most inorganic research involves work with small molecules, and relatively little concentrated effort has been devoted to the macromolecular aspects of the subject. The complexity of the macromolecular chemistry has undoubtedly contributed to this neglect. However, it is clear from recent work that dramatic advances in both fundamental science and technology would be possible if the high polymer chemistry of the representative elements were to be studied in detail. Indeed, the much-heralded renaissance in Main Group chemistry may ultimately depend on a closer investigation of the macromolecular aspects of the field.

My purpose here is to illustrate what can be accomplished with just one inorganic macromolecular system -- in this case constructed from a backbone of phosphorus and nitrogen atoms. Almost certainly, other systems based on the Main Group elements can be developed to an equal or greater degree. I hope that the following comments will stimulate an increased interest in that direction. I will also attempt to illustrate the relationship between the fundamental chemistry and an approach to solving practical problems.

0097–6156/83/0232–0049 $06.00/0

Guiding Principles

Nearly all synthetic polymers are synthesized by the polymerization or copolymerization of different "monomers." The chain growth process may involve the addition chain reactions of unsaturated small molecules, condensation reactions, or ring-opening chain-coupling processes. In conventional polymer chemistry, the synthesis of a new polymer requires the use of a new monomer. This approach is often unsatisfactory for inorganic systems, where relatively few monomers or cyclic oligomers can be induced to polymerize, at least under conditions that have been studied to date. The main exception to this rule is the condensation-type growth that occurs with inorganic di-hydroxy acids.

Because the opportunities for controlled chain growth are more restricted in inorganic than in organic systems, an alternative approach to polymer synthesis becomes appealing. This involves the use of substitution processes carried out on a preformed reactive polymeric intermediate. In this way molecular diversity can be introduced by different substitution reactions rather than by a diversification of the polymerization process.

If this principle can be applied, two potential problems must be avoided: the substitution reactions must lead to neither chain cleavage nor crosslinking.

Simple Substitution Reactions with Poly(dihalophosphazenes)

Poly(dichlorophosphazene) (II) is a highly reactive inorganic macromolecule. It can be prepared by the carefully controlled thermal polymerization of the cyclic trimer, hexachlorocyclotriphosphazene (I), itself synthesized from phosphorus pentachloride and ammonium chloride. In solution, the chlorine atoms in II can be replaced readily by reaction with a wide variety of organic nucleophiles (1,2,3) (Scheme 1). The resultant polymers (III-V) are stable and display a range of physical and chemical properties determined by the nature of the organic side groups. This synthesis process has been reviewed in detail elsewhere (4-7). Here it is sufficient to note that several hundred poly(organophosphazenes) have been prepared by this method. Polymers of this type are already being used in technology; they are also of considerable scientific interest. Similar syntheses have been developed based on poly(difluorophosphazene), $(NPF_2)_n$ (8).

Cyclic Trimers and Tetramers as Reaction Models

From a theoretical and mechanistic point of view, small molecule rings are much easier to study than long macromolecular chains. Substitution reactions carried out on macromolecular

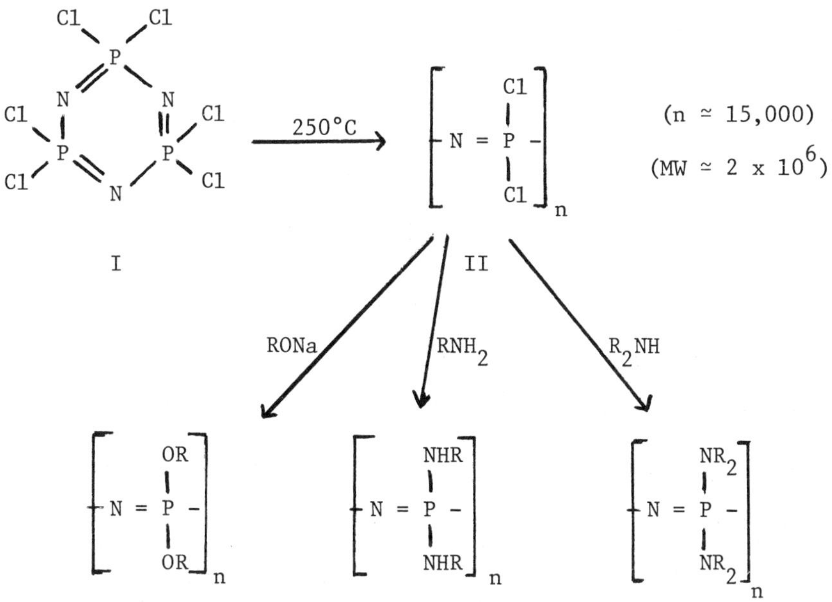

I II III IV V

Scheme 1

substrates may involve side reactions that lead to chain cleavage or crosslinking. Mechanistic studies with macromolecules are difficult to carry out because of solution viscosity effects, distributions in chain length, and the problems of characterization. Hence, it is prudent to explore potential new reactions first with the use of small molecule models such as I, VI, and VII and then to extend these reactions to the high polymers.

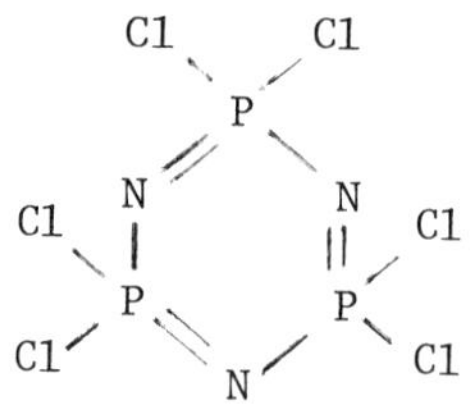

I

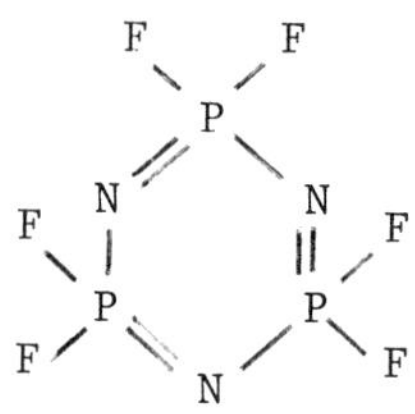

VI

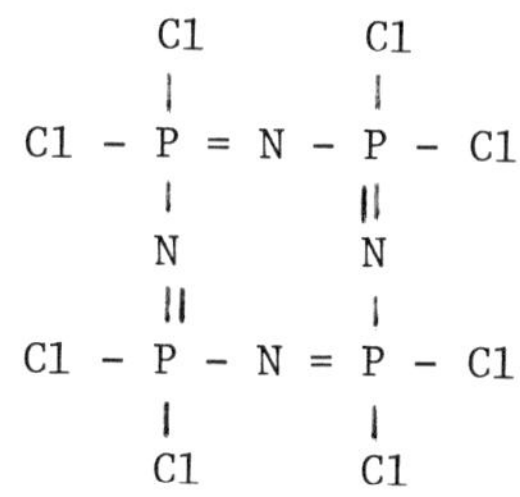

VII

Some of the polymeric reactions mentioned below are still under study at the model compound level.

Modern Objectives in Polymer Synthesis

Polymers have been valued since antiquity for their solid state properties. By this is meant their ability to undergo chain entanglement or co-linear orientation and microcrystallization in the solid state. This underlies their use as structural materials, films, fibers, and elastomers. Such properties still constitute the driving force for most polymer-oriented research, especially with respect to the synthesis of heat-stable, radiation-stable, or highly flexible materials. The electrical properties of solid polymers have always been of interest.

However, in recent years another approach to polymer chemistry has received increased emphasis. In this, macromolecules are studied in terms of their behavior as single molecules rather than as molecular conglomerates. In solution, polymer molecules behave differently from small molecules because the long chain length permits extensive coiling, reduced translational mobility, and an inability to pass through semipermeable membranes. Lightly crosslinked polymers behave like linear polymers in solution except that the swollen matrix has a physical immobility and an open matrix character unlike any other system. For these reasons, polymers are of great interest as "carrier molecules" for chemotherapeutic drugs or transition metal catalysts.

Finally, single macromolecules, because of their one-dimensional character, offer the promise of sequential side group coding, information storage, and template function in the manner that is well known in biological polymers (Figure 1).

Conventional synthetic organic polymers are being studied for all of these reasons, but the general lack of chemical reactivity in these systems is a serious drawback. It is for this reason that polyphosphazenes, with their substitutive

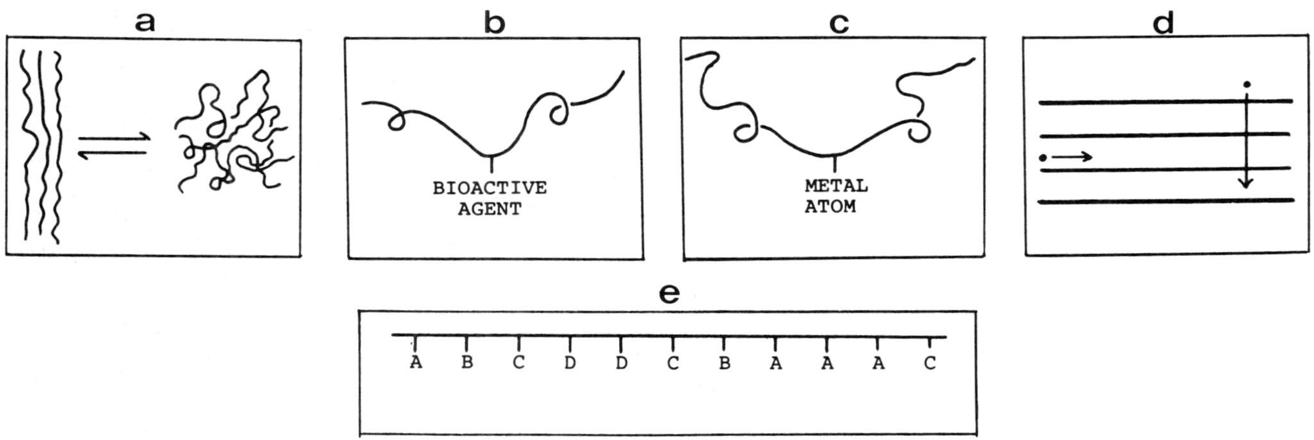

Figure 1. Traditional and exploratory uses for polymers.

Key: a, traditional use of the solid-state, chain entanglement behavior that gives rise to strength or elasticity; b, exploratory use as carrier molecules for bioactive agents in controlled-release drug therapy either as targeted macromolecular drugs or as immobilized biodegradable systems; c, use as immobilizing agent for transition metal catalysts for easier manipulation or recovery or to modify the catalytic activity; d, use as electrical conductors or semiconductors (conduction may occur along unsaturated chain sequences or between chains in crystalline domains); e, sequential arrangement of side groups along a linear polymer chain offers the prospect of (1) control of polymer conformation; (2) use as a template for the controlled construction of complementary polymer molecules; and (3) information storage at the molecular level.

method of synthesis, are of considerable interest. In the following sections, I will illustrate why the polyphosphazene system is an appealing starting point for new developments in two specific areas -- in chemotherapy and polymer-bound catalyst work.

Bioactive Polyphosphazenes

Specific inorganic macromolecules are unusual because they can be hydrolyzed to relatively innocuous products or to small molecules that can be metabolized. Most conventional organic polymers do not have this attribute. Thus, these inorganic systems are of special interest as carrier molecules in chemotherapy.

Recent work in our laboratory has shown that certain side groups attached to a polyphosphazene chain impart a sensitivity to hydrolytic chain cleavage; other side groups generate water-solubility. Both of these characteristics are important in chemotherapy. Polyphosphazenes are also valuable in biology because two or more substituted groups can be readily attached to the same chain. Thus, individual side groups that possess chemotherapeutic, water-solubilization, hydrolytic-destabilization, or "homing" characteristics can be combined in one molecule to form a drug with a set of synergistic properties.

Polymers containing the repeating units shown in VIII-XI have been shown to be hydrolytically degradable and/or water-soluble (9-13). Amino acid ester derivatives (VIII) degrade to ethanol, amino acid, phosphate, and ammonia, which can either be metabolized or excreted. Thus, such side units used together with chemotherapeutic cosubstituent groups, provide a facile drug delivery system. Imidazolyl side groups (IX) also confer hydrolytic sensitivity, but the biochemical response to the hydrolysis products has not yet been established. Methylamino side groups (X) provide water-solubility, as do glucose residues (XI).

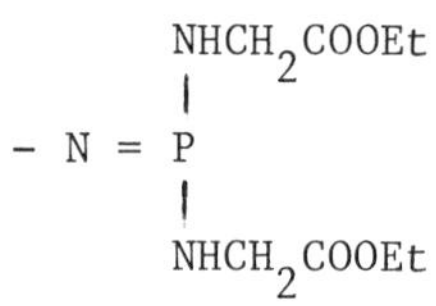

VIII

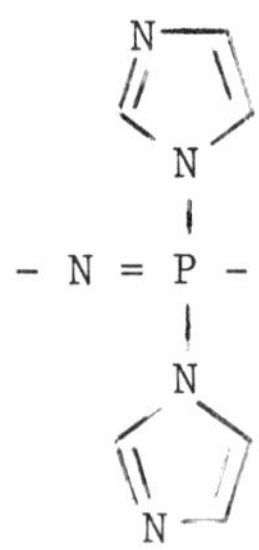

IX

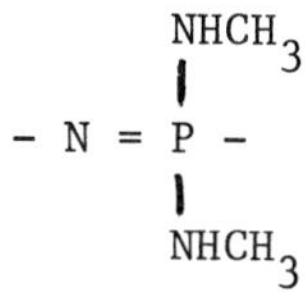

X

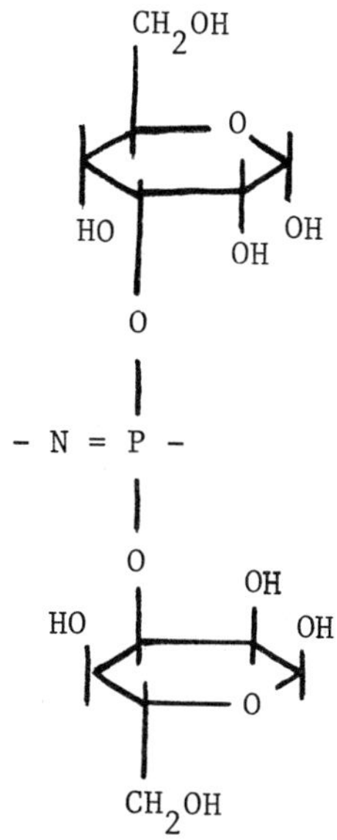

XI

These polymers were synthesized by the general methods shown in Scheme 1. Their hydrolysis behavior has been the subject of several fundamental mechanistic studies at the model compound level (10,11).

The attachment of biologically-active side groups has also been explored. At the present time several different approaches have been developed which lead to the synthesis of polymers such as XII-XVII.

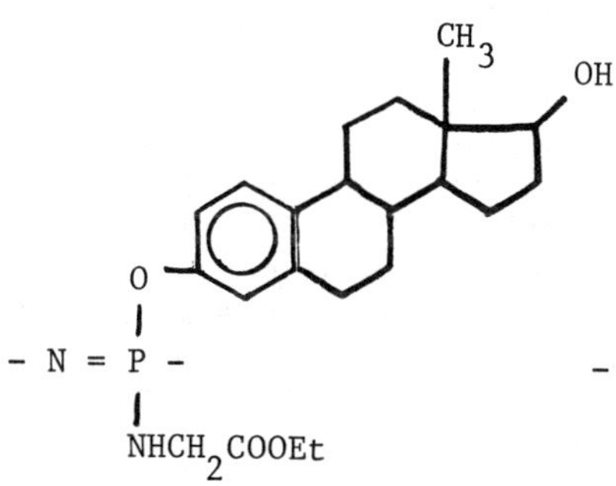

XII

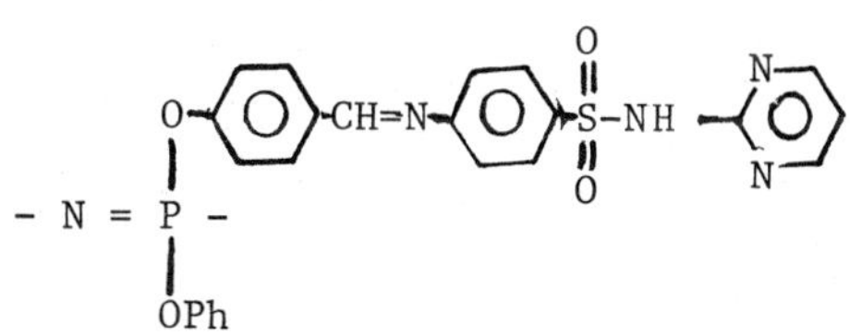

XIII

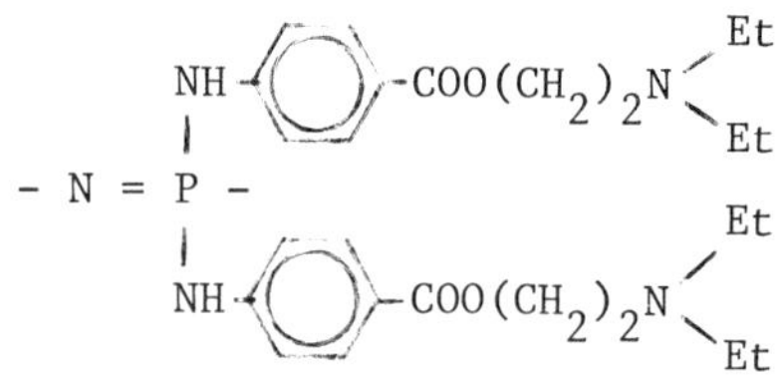

XIV

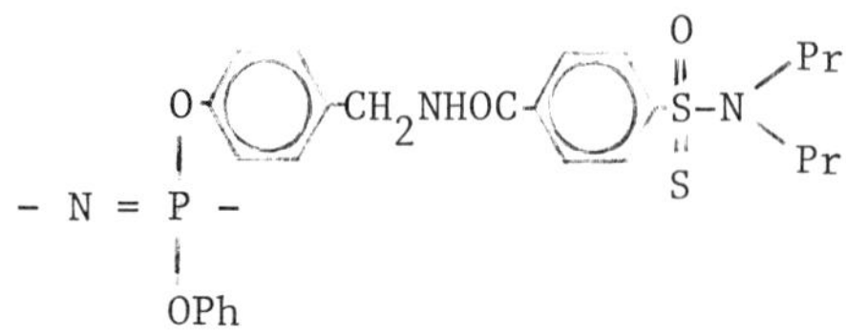

XV

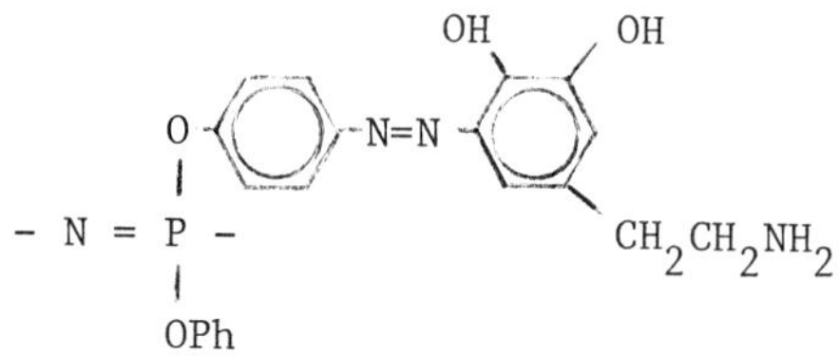

XVI

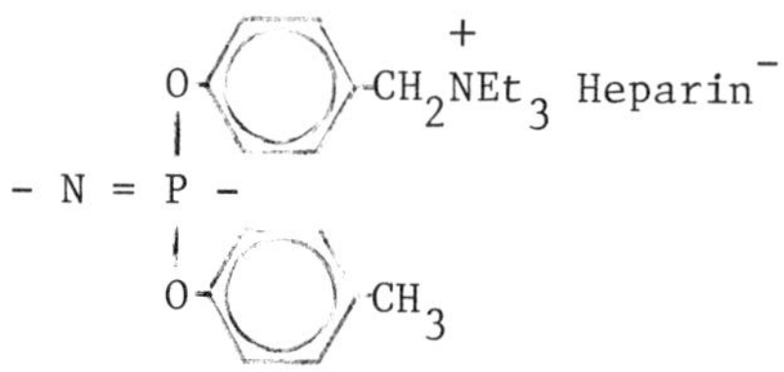

XVII

Steroid-bound polyphosphazenes (XII) can be prepared by the reaction of II with metal steroidoxide salts followed by treatment with amino acid esters (14,15). The sulfadiazine-bound polymer (XIII) was synthesized by Schiff's base coupling of a polyphosphazene bearing a pendent aldehydic group with the antiobiotic amine (16). The local anesthetic, procaine, was linked to the polymer backbone by direct aminolysis of II to yield polymers based on the repeating unit XIV (17). Peptide-coupling techniques have been used for the linkage of polyphosphazenes bearing pendent amino residues to bioactive carboxylic acids (XV) (18). Catecholamines, such as dopamine or epinephrine, have been linked to aryloxyphosphazene high polymers by diazo coupling methods (XVI) (19), and these polymers retain the biological activity of the free hormone. The mucopolysaccharide, heparin, can be bound to aryloxyphosphazene polymers via ionic exchange with quaternized ammonium pendent groups (XVII) (20). These polymers show promise as non-thrombogenic materials for bio-engineering.

It will be clear that, combined with the use of hydrolytically-sensitizing or water-solubilizing cosubstituent groups, these polymers could have an important impact on chemotherapy and other areas of biomedicine. At present, the problem is to establish the feasibility of a wide range of synthetic methods and to evaluate the biological activity of each class of polymers.

Linkage of Transition Metals to Phosphazene Rings and High Polymers

This book comprises a survey of recent advances in the chemistry of the representative (Main Group) elements. However, the long-range resurgence of Main Group chemistry as an expanding research area depends to some extent on the strength of its interface with organic chemistry, transition metal chemistry, and applied science. Some organic-related aspects of phosphazene chemistry were discussed in the previous section. Here the interface with transition metal chemistry is reviewed.

Polyphosphazenes and cyclophosphazenes are almost unique as carrier molecules for transition metals because of the wide range of binding sites that can be incorporated into the phosphazene structure. The substitutive mode of synthesis described earlier allows a structural diversity that is not found, for example, in polystyrene, polyphenylene oxide, or other organic carrier polymers.

The emphasis in the following sections will be on exploratory model reactions carried out with phosphazene cyclic trimers or tetramers, although the analogous macromolecules systems have also been studied in several cases. First, I will summarize the various types of metal binding sites that are accessible at the present time. Synthetic procedures leading to the incorporation of several of these sites and their role in metal binding will then be discussed.

Options for Metal-Binding Sites. Seven approaches for metal-binding to cyclic or polymeric phosphazenes have been explored in our laboratory. These are summarized in structures XVIII (21), XIX (22-25), XX (26), XXI (27), XXII (28,29), XXIII (30), and XXIV (31,32,33).
Only three of these approaches will be discussed here. The others can be traced through the references given.

Pendent Phosphine Groups. The classical method for the linkage of transition metal units to high polymers is via pendent

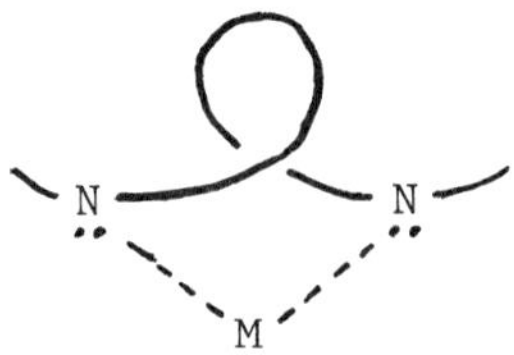

XVIII

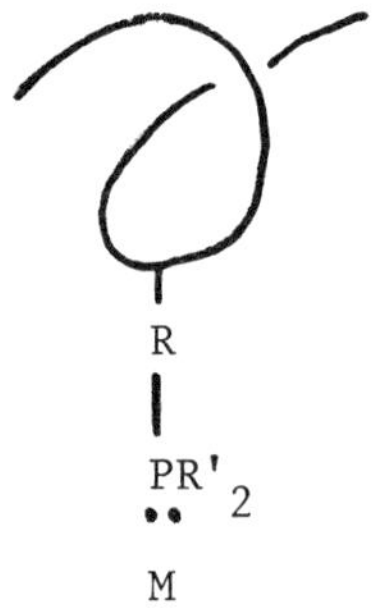

XIX

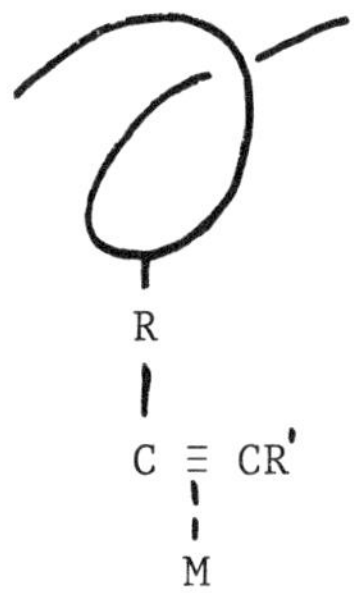

XX

XXI

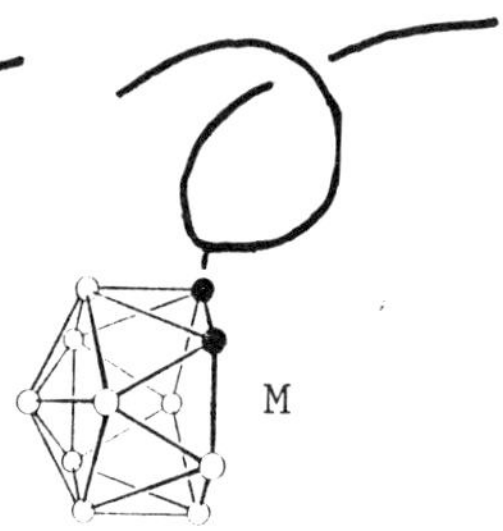

XXII

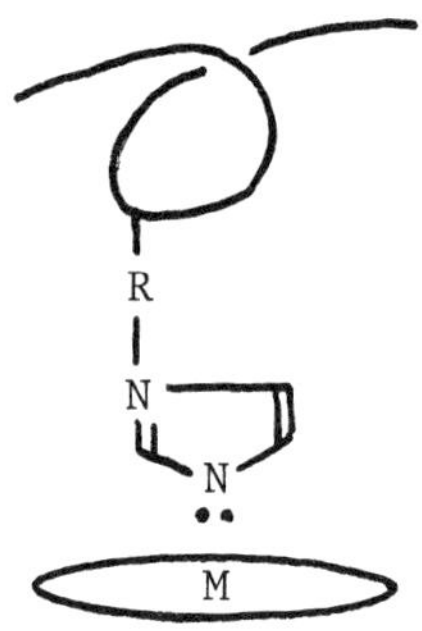

XXIII

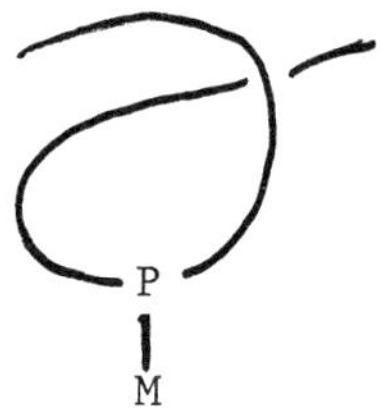

XXIV

phosphine groups attached to a macromolecular chain. We have developed a synthetic strategy for the preparation of cyclophosphazene model compounds and the corresponding linear high polymers which bear pendent triarylphosphine groups. This approach is illustrated in Scheme 2 (22-25).

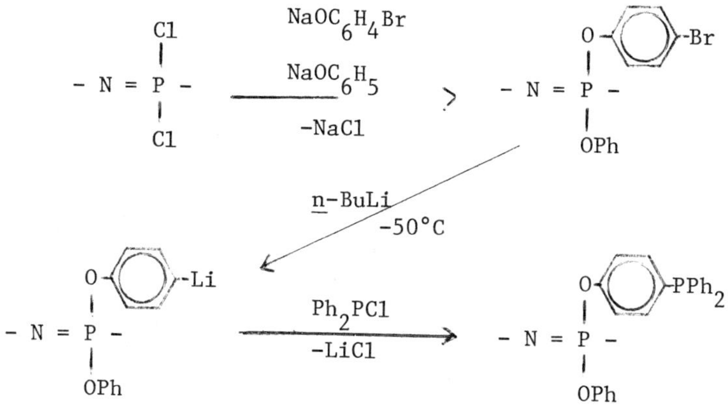

Scheme 2

Typical products prepared by this route include XXV, XXVI, and XXVII.

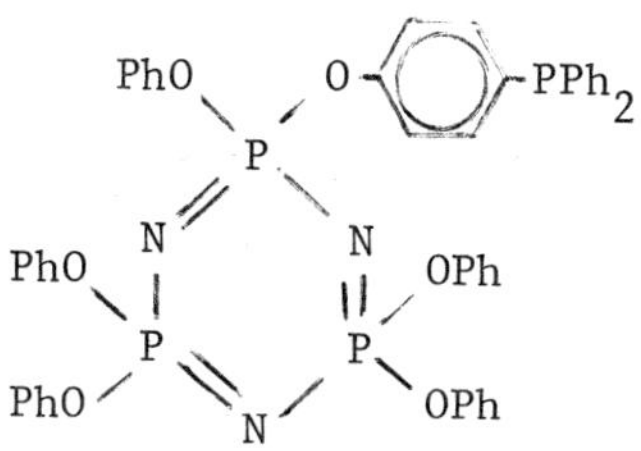

XXV

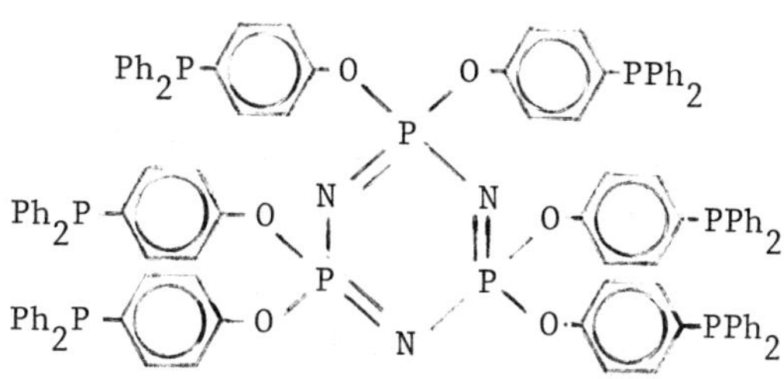

XXVI

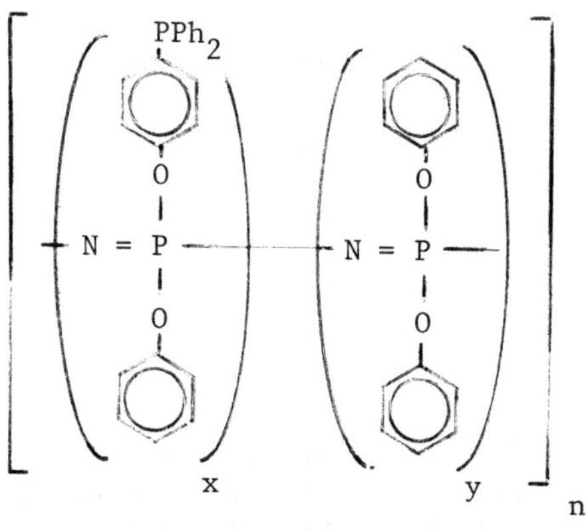

(n ≃ 7,000)

XXVII

Species such as XXV, XXVI, or XXVII readily form coordination complexes when treated with AuCl, $H_2Os_3(CO)_{10}$, $Mn(CO)_3(\eta\text{-}C_5H_5)$, $Fe(CO)_3(PhCH{=}CHC(O)CH_3)$, or $[RhCl(CO)_2]_2$ (25). Two results are of special interest. First, the skeletal nitrogen atoms in XXV-XXVII do not participate in the coordination process. Presumably, they are effectively shielded by the aryloxy units and are of low basicity. Second, coordinative crosslinking can occur when two phosphine residues bind to one metal atom. Ligand-exchange reactions were detected for the rhodium-bound species. The tri-osmium cluster adducts of XXV, XXVI, and XXVII are catalysts for the isomerization of 1-hexane to 2-hexene.

Carboranyl Phosphazenes. Cyclic trimeric and high polymeric chlorophosphazenes react with lithiocarboranes to form carboranyl phosphazenes, as shown in XXVIII and XXIX (28).

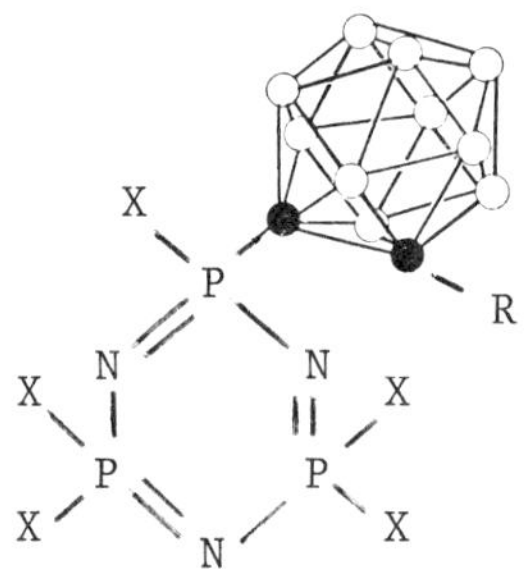

XXVIII

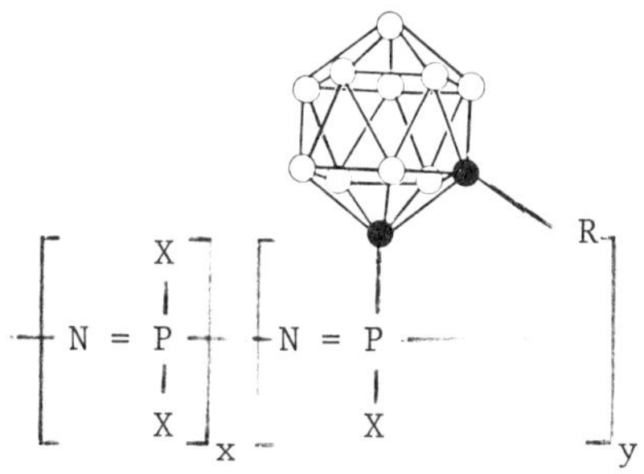

XXIX

(O = BH)

R = Me or Ph, and X = Cl or OCH_2CF_3 (● = C)

The halogen atoms remaining can then be replaced by organic residues such as trifluoroethoxy units. High polymers can also be prepared by ring-opening polymerization of the chlorocyclophosphazene, XXVIII. Compounds of this type can be converted to nido-carboranes in the presence of base, but these do not form metallo-derivatives, presumably for steric reasons (29).

However, separation of the carborane cage from the phosphazene ring or chain by a methylene spacer group allows metals to be inserted into the open face of the carborane. These syntheses were accomplished by the reaction routes shown in Schemes 3 and 4. High polymeric analogues of these transformations have also been accomplished following polymerization of XXX. The rhodium-bound cyclophosphazenes and polyphosphazenes are catalysts for the hydrogenation of 1-hexene. In this, they show a similar behavior to metallocarboranes linked to polystyrene

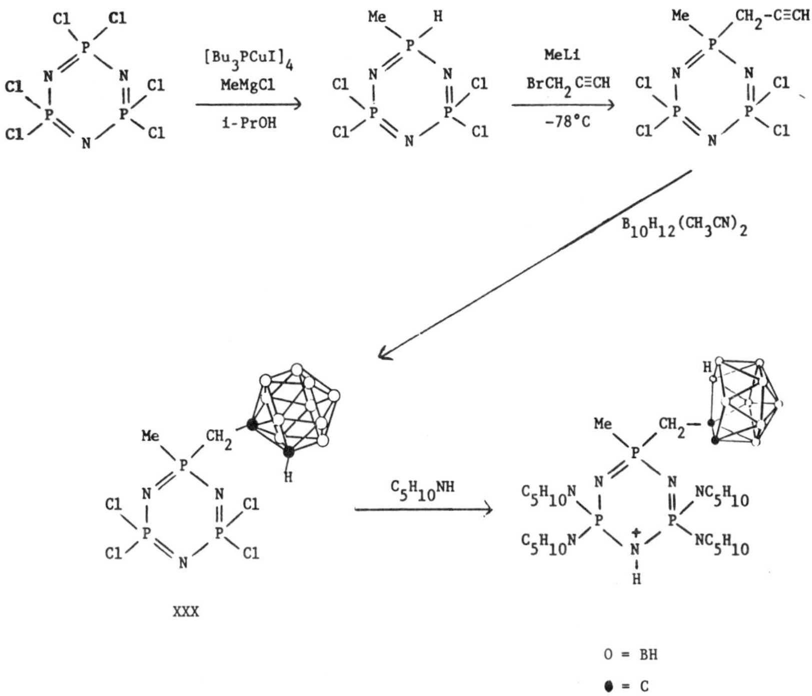
[Bu3PCuI]4
MeMgCl
i-PrOH
MeLi
BrCH2C≡CH
-78°C
B10H12(CH3CN)2
C5H10NH
XXX
O = BH
● = C

Scheme 3

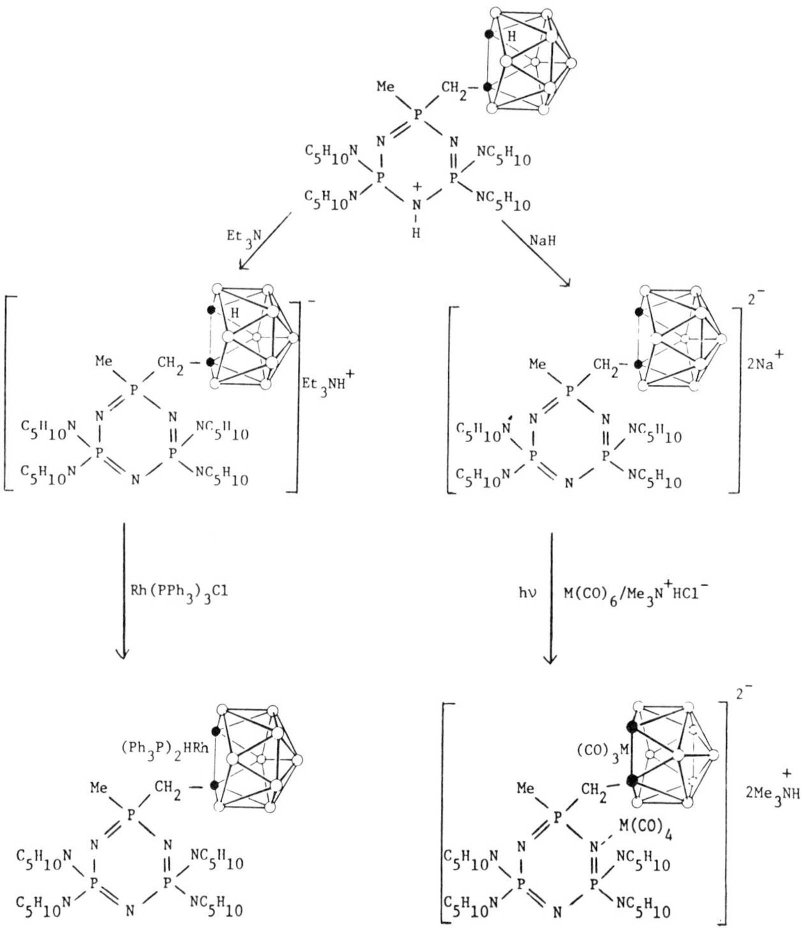
Me
CH_2
H
$C_5H_{10}N$
NC_5H_{10}
N
P
H
Et_3N
NaH
Et_3NH^+
$2Na^+$
$Rh(PPh_3)_3Cl$
hν
$M(CO)_6/Me_3N^+HCl^-$
$(Ph_3P)_2HRh$
$(CO)_3M$
$M(CO)_4$
$2Me_3NH^+$
2^-

Scheme 4

(34). The oxidative-stability of the phosphazene backbone is expected to be an advantage in catalytic reactions.

Metallophosphazenes with Phosphorus-Metal Bonds. Until recently, the chemistry of cyclic and high polymeric phosphazenes was essentially the chemistry of their organic derivatives (Scheme 1). However, a discovery reported in 1979 (31) opened up a new field of metallophosphazene chemistry in which transition metals form the nucleus of the side group structure and are linked to the skeleton through phosphorus-metal bonds. These species are of theoretical and potentially practical importance, and I will summarize briefly some of the main features known at this time.

Organometallic anions react with halophosphazenes to replace halogen atoms by organometallic units. The first reactions of this type discovered are illustrated in Scheme 5. The metallophosphazenes are surprisingly stable. Moreover, as shown below, hexachlorocyclotriphosphazene reacts with an organometallic *di*-anion to yield both a dimetallo derivative (XXXI) and a trimetallic cluster derivative (XXXII). The latter compound is

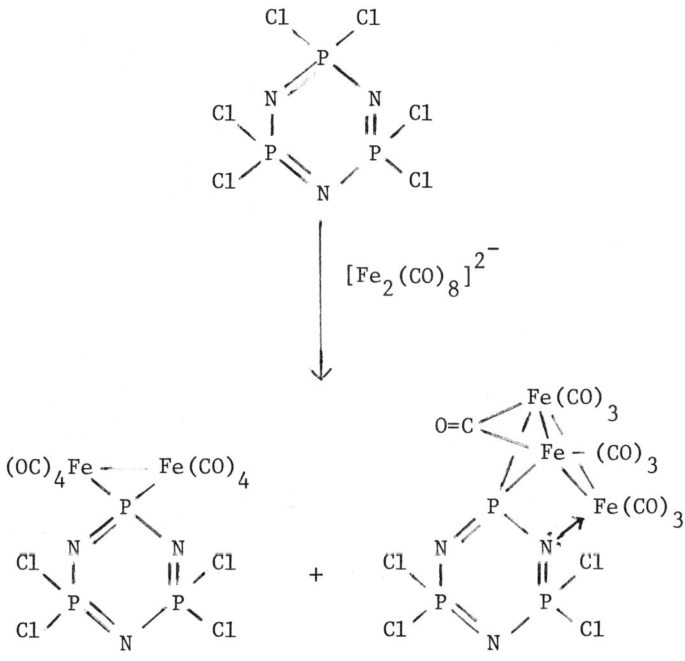

XXXI XXXII

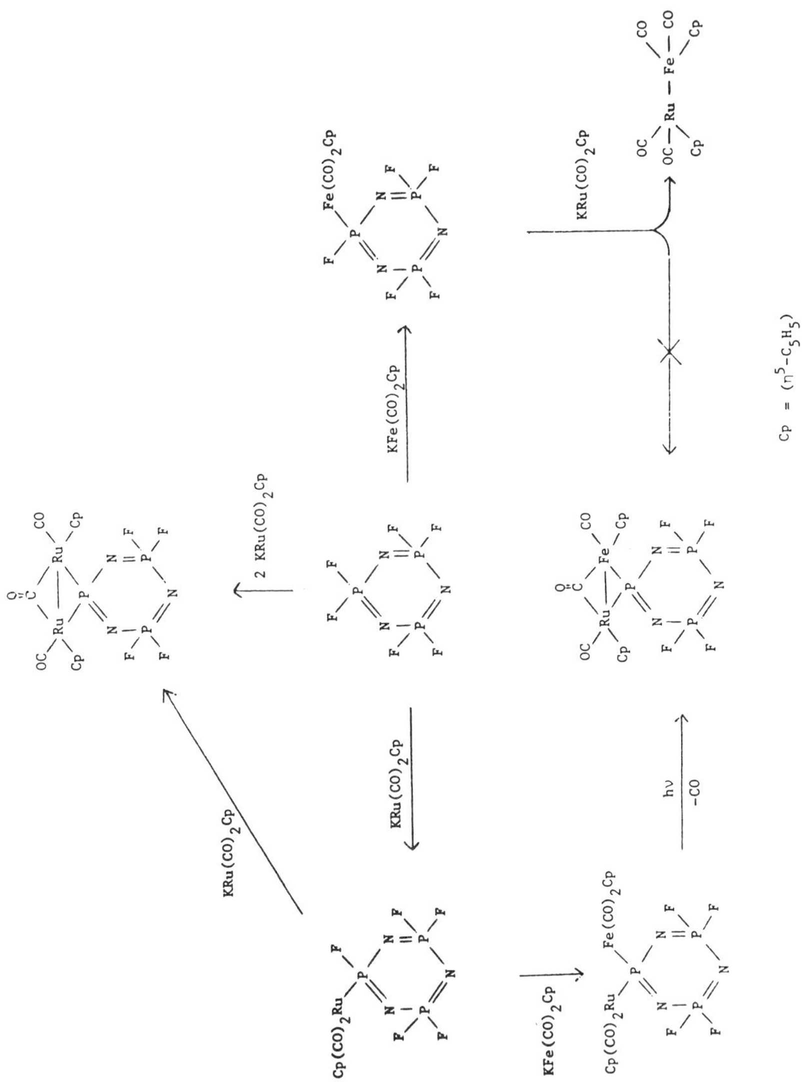
2 KRu(CO)2Cp
KFe(CO)2Cp
KRu(CO)2Cp
KRu(CO)2Cp
KRu(CO)2Cp
KFe(CO)2Cp
hν
-CO
Cp = (η5-C5H5)

Scheme 5

stabilized by both P-metal and N-metal coordinative bonds. Other metallophosphazenes containing Pt and Au have been reported recently by Schmidpeter and coworkers (35). X-ray structure data have been obtained for several of these compounds and, as might be expected, a strong interaction between the metallic units and the phosphazene ring system is evident (31,32,33,36).

Conclusions

The reactions discussed in this chapter are illustrative of the chemical diversity that follows from the substitutive approach to polymer synthesis. Rings and polymers based on the Main Group inorganic elements are especially appropriate for this approach because of the generally high reactivity of, for example, the element-halogen bonds. Thus, a key problem facing the inorganic research community is to devise and develop methods for the synthesis of high polymers, comparable to poly-(halophosphazenes), that contain elements such as silicon, aluminum, boron, or sulfur in the skeleton, with reactive side groups attached to these atoms. Once these polymers have been synthesized, a diverse arsenal of side group substitution processes can then be mobilized to prepare a broad range of different macromolecules. If this can be accomplished, it should have an almost unprecedented impact on inorganic chemistry, polymer chemistry, and high technology.

Acknowledgments

The work described in this chapter was funded by the U.S. Army Research Office, the Office of Naval Research, the National Institutes of Health, NASA, and the National Science Foundation Polymers Program. It is a pleasure to acknowledge the contributions of my recent coworkers, especially T. J. Fuller, K. Matsumura, T. L. Evans, P. E. Austin, T. X. Neenan, K. D. Lavin, N. M. Tollefson, G. H. Riding, L. T. Wagner, P. R. Suszko, M. S. Connolly, J. L. Desorcie, P. J. Harris, A. G. Scopelianos, M. L. Levin, and R. J. Ritchie. Those coworkers who contributed to the earlier work are cited in the list of references.

Literature Cited

1. Allcock, H. R.; Kugel, R. L. J. Am. Chem. Soc. 1965, 87, 4216.
2. Allcock, H. R.; Kugel, R. L.; Valan, K. J. Inorg. Chem. 1966, 5, 1709.
3. Allcock, H. R.; Kugel, R. L. Inorg. Chem. 1966, 5, 1716.
4. Allcock, "Phosphorus-Nitrogen Compounds"; Academic Press: New York, 1972; Chapters 15 and 16.
5. Allcock, H. R. Makromol. Chem. (Proc. IUPAC Conf. on Macromolecules, Florence, Italy, 1980), 1981, Suppl. 4, 3.

6. Singler, R. E.; Hagnauer, G. L.; Sicka, R. W. ACS. Symp. Ser. 1982, p.11.
7. Tate, D. P. J. Polym. Sci., Polym. Symp. 1974, 48, 33.
8. Evans, T. L.; Allcock, H. R. J. Macromol. Sci. - Chem. 1981, A16, 409.
9. Allcock, H. R.; Fuller, T. J.; Mack, D. P.; Matsumura, K.; Smeltz, K. M. Macromolecules. 1977, 10, 824.
10. Allcock, H. R.; Fuller, T. J.; Matsumura, K. Inorg. Chem. 1982, 21, 515.
11. Allcock, H. R.; Fuller, T. J. J. Am. Chem. Soc. 1981, 103, 2250.
12. Allcock, H. R.; Cook, W. J.; Mack, D. P. Inorg. Chem. 1972, 11, 2584.
13. Allcock, H. R.; Scopelianos, A. G. Macromolecules. 1983, in press.
14. Allcock, H. R.; Fuller, T. J.; Matsumura, K. J. Org. Chem. 1981, 46, 13.
15. Allcock, H. R.; Fuller, T. J. Macromolecules. 1980, 13, 1338.
16. Allcock, H. R.; Austin, P. E. Macromolecules. 1981, 14, 1616.
17. Allcock, H. R.; Austin, P. E. Macromolecules. 1982, 15, 680.
18. Allcock, H. R.; Neenan, T. X.; Kossa, W. C. Macromolecules. 1982, 15, 693.
19. Allcock, H. R.; Hymer, W. C.; Austin, P. E. Macromolecules. 1983 (in press).
20. Neenan, T. X.; Allcock, H. R. Biomaterials. 1982, 3, 78.
21. Allcock, H. R.; Allen, R. W.; O'Brien, J. P. J. Am. Chem. Soc. 1977, 99, 3984.
22. Evans, T. L.; Fuller, T. J.; Allcock, H. R. J. Am. Chem. Soc. 1979, 101, 242.
23. Allcock, H. R.; Evans, T. L.; Fuller, T. J. Inorg. Chem. 1980, 19, 1026.
24. Allcock, H. R.; Fuller, T. J.; Evans, T. L. Macromolecules. 1980, 13, 1325.
25. Allcock, H. R.; Lavin, K. D.; Tollefson, N. M. Organometallics. 1983, 2, 267.
26. Harris, P. J.; Nissan, R. A.; Allcock, H. R. J. Am. Chem. Soc. 1981, 103, 2256.
27. Suszko, P. R.; Whittle, R. R.; Allcock, H. R. J. Chem. Soc. Chem. Commun. 1982, 960.
28. Allcock, H. R.; Scopelianos, A. G. O'Brien, J. P.; Bernheim, M. Y. J. Am. Chem. Soc. 1981, 103, 350.
29. Allcock, H. R.; Scopelianos, A. G.; Tollefson, N. M.; Whittle, R. R. J. Am. Chem. Soc. 1983, 105, 1316.
30. Allcock, H. R.; Greigger, P. P.; Gardner, J. E.; Schmutz, J. L. J. Am. Chem. Soc. 1979, 101, 606.
31. Greigger, P. P.; Allcock, H. R. J. Am. Chem. Soc. 1979, 101, 2492.
32. Allcock, H. R.; Greigger, P. P.; Wagner, L. J.; Bernheim, M. Y. Inorg. Chem. 1981, 20, 716.

33. Allcock, H. R.; Wagner, L. J.; Levin, M. L. J. Am. Chem. Soc. 1983, 105, 1316.
34. Paxson, T. E.; Hawthorne, M. F. J. Am. Chem. Soc. 1974, 96, 4674.
35. Schmidpeter, A.; Blank, K.; Riffel, H. Angew. Chem. 1980, 92.
36. Suszko, P. R.; Whittle, R. R.; Allcock, H. R. J. Chem. Soc. Chem. Commun. 1982, 960.

RECEIVED April 27, 1983

4

Polyanionic Clusters in Solids and Their Chemical Reactions

H. G. VON SCHNERING

Max-Planck-Institut für Festkörperforschung Heisenbergstr. 1, D-7000 Stuttgart 80, Federal Republic of Germany

Structural units of binary polycompounds with main group elements consist of one-, two- and three-dimensional polymers, as well as cage-like polycyclic anions. These polyanions can be converted into molecular compounds by appropriate chemical reactions. The reactions, which are discussed in general, clearly provide an easy means of preparation for new types of compounds. Compounds such as P_7R_3, As_7R_3 and $P_{11}R_3$ have been obtained by these methods. Further homogeneous reactions with P_7R_3 yield other $P_7R'_3$ (R'=Sn me$_3$, Pb me$_3$) compounds which are surprisingly stable. Reactions of Rb_3As_7, for example, in ethylendiamine produce $Rb_3As_7\cdot3en$, which serves as a model of the structure in solution. Redox reactions of Na_3P_7 or Na_3P_{11} in THF yield P_{16}^{2-} and $Na_4P_{14}\cdot6en$ which can be seen as the first steps of polycondensation of the P_7 units. The formation of allo-Ge and 4H-Ge from Li_7Ge_{12} is an especially interesting reaction in general, in view of the preparation of new (metastable) modifications of the elements. Some of the binary polycompounds possess interesting physical properties. The congruent sublimation, the crystalline to plastically crystalline phase transitions as well as the raman spectra are discussed in detail.

The electropositive elements form series of solid binary compounds MY_n with the main group 3B, 4B, 5B, 6B elements. Independent of the real bonding present in the solids, the structures can be described in terms of homoatomic polyanionic cluster formation, Y_n^{x-}. Such polyanionic clusters exist as one-, two- and three-dimensional connected polymers, as well as quasi isolated cage-like polycyclic units. Some examples are given in Fig. 1 and a more general discussion was published recently (1).

0097–6156/83/0232–0069 $06.00/0

The following contribution essentially deals with the utilization of these solid binary compounds as sources for new polycyclic molecular systems or for new metastable modifications of the elements. New results are discussed which contribute to a general understanding of these materials and their properties.

Metal Compounds as Sources for Polycyclic Systems

The winning of molecular compounds from heterogeneous reactions of solid materials is a well known process (eg. C_2H_2 from CaC_2). A general scheme is given in Figure 2 for compounds of the type M_3Y_7 and M_3Y_{11}, respectively (M = alkali metal; Y = P, As, Sb). In the structures of these compounds cage-like Y_7^{3-} or Y_{11}^{3-} anions are present; thus reactions to neutral molecules of the same structure suggest themselves. The given scheme however, can also be generalized for application to polymeric polyanions.

Heterogeneous Reactions. In a heterogeneous reaction of solid $M_3Y_{7,c}$ with suitable reactants RX (eg. $ClSime_3$) solutions of $R_3Y_{7,sol}$ are obtained, from which the molecular crystalline compounds $R_3Y_{7,c}$ result (2,3). A direct path via substituent exchange then leads from these materials to new compounds, often with high yields (4). Several examples are given in Figure 3: $As_7(Sime_3)_3$; $P_{11}(Sime_3)_3$; $P_{11}(Snme_3)_3$; and above all the complete series $P_7(Mme_3)_3$, with M = Si, Ge, Sn, Pb as well as $P_7(Mph_3)_3$ with M = Sn, Ge (5,6,7).

In the above mentioned series, surprisingly the stability of the compounds in air increases. The silicon compounds decompose instantaneously into phosphane and siloxane; the lead compounds in contrast, however, react only in the course of some days. The systematic changes in the i.r. spectra are also interesting. The M-P and M-C vibrations show the expected shift with atomic mass, whereas the vibrations of the P_7-cages remain completely unaffected by the substitution (4).

All attempts to bond R_3P_7 as a ligand to metal(0) complexes have failed. In contrast, the preparation of complexes with the related cage-molecules $P_4(Sime_2)_3$ was successful (8).

Homogeneous Reactions. With suitable solvents the binary metal compounds yield clear, homogeneous solutions. Especially suitable are ethylendiamine (en), liquid NH_3, and in some cases THF. The deep red solutions in en have been known for quite a long time (9), and seem to be typical for the great majority of compounds with polyanions (cf. the solution of Sn_9^{4-} (10,11,12)). We now observe, that solutions of pure binary polyphosphides, eg. Na_3P_7, in very carefully purified ethylendiamine (distilled over $LiAlH_4$; controlled by conductivity measurements; specific conductivity lower by 10^{-1} than that given in the literature) are absolutely colourless. The well-known deep red colour again appears instanteneously with traces of oxygen (produced by the

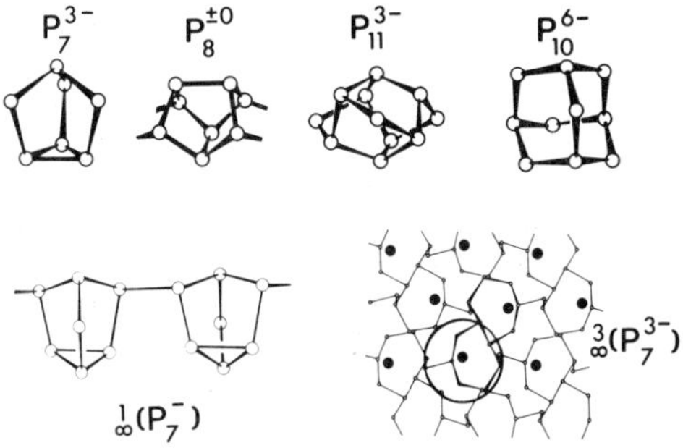

Figure 1. Some homoatomic phosphorus units in metal polyphosphides as examples for polyanionic clusters.

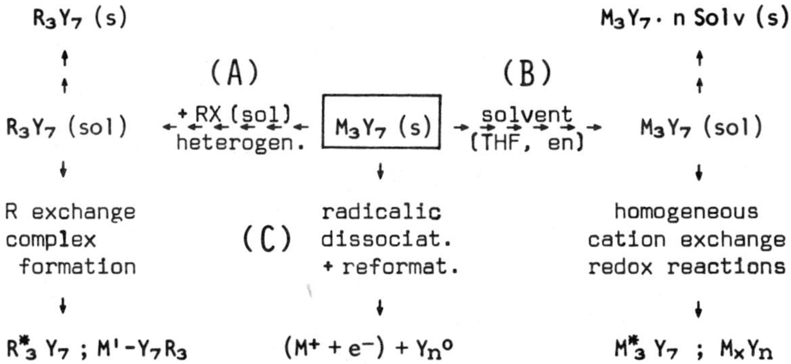

Figure 2. Different reaction paths A, B, and C of a solid polyanionic cluster shown by the example of M_3Y_7 (s).

decomposition of HgO). Another observation concerns white phosphorus: pure P_4 dissolves in en giving a deep red colour; the solution is discoloured by oxygen. We suspect that radical species are responsible for the intense red colour and investigations are in progress.

Considerable uncertainty still exists about the state of the metal compounds in the solutions: "It's dark behind the pick" (old miner's saying)! M. Baudler assumed the presence of Li^+ and P_7^{3-} ions for Li_3P_7 in THF solution (13). In these solutions valence fluctuations of the P_7^{3-} cage were observed for the first time (14). The details were explained by Gleiter (15) by a theoretical model. The valence fluctuations freeze at 240 K (^{31}P-NMR). We were recently able to prove the fluctuations for Na_3P_7 in en (5). After the structure determination of solvated compounds, some doubts exist concerning the path of this valence fluctuation. The cations in the crystalline solvates $Rb_3As_7\cdot3en$ and $Na_4P_{14}\cdot6en$ are bonded exclusively as in an undissociated ion-complex to only one anion and are coordinated only by solvate molecules (Figure 4) (16). The unusual configuration corresponds to an isolated solvated M_3Y_7 molecule and is quite different than the configuration in solvent-free Li_3P_7 (6) and Na_3P_7 (17). We also regard this structure as a good model for dissolved compounds and believe that a Metal Supported Valence Fluctuation (MSVF) takes place in the solution. A simple electrostatic estimation shows that the participation of the metal atoms noticeably decreases the energy difference between ground-state and transition state (16).

The solutions of the metal compounds M_3X_7 can be used (a) for a study of redox-reactions under mild conditions, and (b) to serve as aducts for cation exchange reactions. Thus Na_3P_7 reacts with ph_4PCl or ph_4AsCl accompanied by quantitative precipitation of NaCl, to give solutions of salts without metal cations. The least soluble materials which crystallize out of these solutions are the compounds $(ph_4P)_2P_{16}$ or $(ph_4As)_2P_{16}$. They contain the new polyanion P_{16}^{2-} (Figure 5), which is formed in the solution by a complex redox reaction (18).

Na_4P_{14} is formed quantitatively from Na_3P_7 and P_4 (in en) in a well defined reaction. The colourless solvate $Na_4P_{14}\cdot6en$ can be obtained from en in crystalline form by counter-current diffusion of THF (Figure 5). Again we find a solvated molecule or ion-complex Na_4P_{14} in the crystal. The polyanion P_{14}^{4-}, whose trimethylsilyl substituted derivatives have already been observed in small quantities in mass-spectra (8), constitutes the first methodically executed step to an oligomerization and finally to a complete polymerization of P_7^{3-} clusters (16).

Radicalic Decomposition. A third path for reactions of the binary metal compounds being discussed here is radicalic dissociation under the influence of suitable reactants. The thermal decomposition into the elements is always possible at high temperatures. At 298 K however, it can only take place if at least one

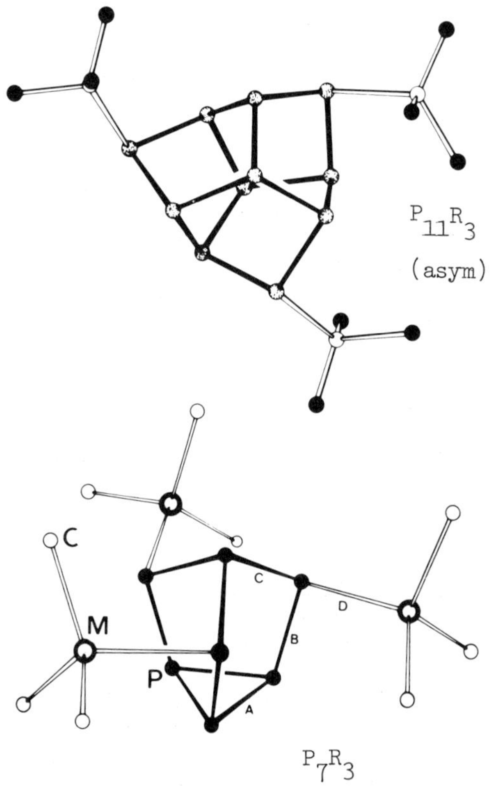

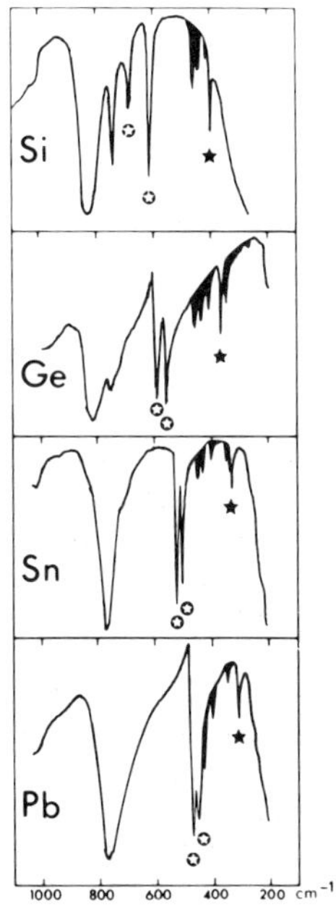

Bond distances

M =	Si	Ge	Sn	Pb	
A =	2.214	2.234	2.226	2.212	Å
B =	2.192	2.195	2.189	2.176	Å
C =	2.180	2.184	2.180	2.178	Å
D =	2.288	2.355	2.539	2.611	Å

Figure 3. Structures of $P_{11}R_3$ and P_7R_3 as examples for substituted polyanions. The IR spectra of the compounds $P_7(Mme_3)_3$ with M = Si, Ge, Sn, and Pb show the nearly unchanged fundamental vibrations of the P_7 skeleton (black-filled), the shifts of $\nu(P\text{–}M)$ (black star), and of $\gamma(M\text{–}C_3)$ (white star), and the CH_3 rocking vibrations (not marked).

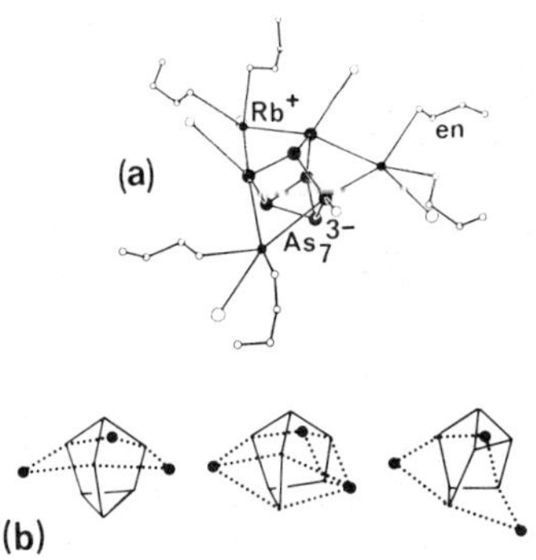

Figure 4. Structure of $Rb_3As_7 \cdot 3en$ (a) showing the solvated ion complex $(Rb^+)_3As_7^{3-}$ with only weak additional bonds between neighboring complexes, and the proposed Metal-Supported Valence-Fluctuation (MSVF) (b) for the X_7^{3-} cage.

of the components can react to give stable products. This is possible in the case of compounds such as Na_3P_7 if Na^o=(Na^++e^-) is transformed into Na^+,sol and a stable radical anion $(R^{\cdot})^-$. Na_3P_7 does in fact react with a solution of benzophenone ph_2CO in THF in an heterogeneous reaction to yield the well-known deep-blue radical anions $(ph\ CO^{\cdot})^-$ and dissolved Na^+ cations. The polyanion P_7^{3-} is discharged by this reaction and forms a light-red amorphous residue of a new, but not yet characterized form of elementary phosphorus. However, a reaction of Li_7Ge_{12} with ph_2CO in THF has been explained in detail (19) (Figure 6).

Li_7Ge_{12} is a new phase in the Li/Ge system, distinguished by a complicated two-dimensional infinite polyanionic structure of Ge atoms. The compound, which is very reactive towards protic agents, immediately reacts in the above described way to give the radical anion and a new metastable modification of germanium, allo-Ge. An exclusively topotactical reaction is involved here, in which the complex 2-dimensional layers of the original Ge_{12}^{7-} polyanions, which are no longer saturated with respect to valency, polymerize directly to allo-Ge.

Allo-Ge is a diamagnetic semiconductor with mechanical properties similar to graphite. Because of the topotactical reaction path, Li_7Ge_{12} single-crystals are transformed directly into single crystals of allo-Ge. Apart from this, Li_7Ge_{12} reacts with protic agents and also topotactically to Li^+,sol, H_2,g and allo-Ge, but very surprisingly, not by the formation of germanes. Hydrogen appears in this reaction exclusively at the edges of the two-dimensional Ge_{12}^{7-} anions, which corresponds excellently to the given scheme in Figure 6. Allo-germanium possesses another very interesting property. Heating it up to 420 K causes it to change by an exothermic reaction into a further metastable modification of germanium, 4H-Ge. This is one of the stacking variants of tetrahedral structures, which are known from SiC modifications and are of considerable theoretical interest, because here in a tetrahedral structure of 4B-elements, six-membered rings appear both in the chair- and in the boat-conformation. Similar to the well-known high pressure form 2H-Si, the axial ratio c/a of the hexagonal unit cell of 4H-Ge deviates from the theoretical value, an indication of the influence of the unfavourable boat-conformation (ecliptic configuration of a part of the Ge-Ge bonds). A slow transformation of 4H-Ge to the stable modification α-Ge with diamond structure begins at about 770 K. Rapid heating will cause 4H-Ge to melt, the melting point being about 10° below the melting point of α-Ge (1198 K/1208 K) (19,20).

These reactions yielding metastable new forms of elements should be studied further.

Crystalline and Plastic Phases M_3P_7 and M_3P_{11} (M=alkaline metal)

To make the above mentioned reactions successful, well defined starting materials are required. By a series of very extensive

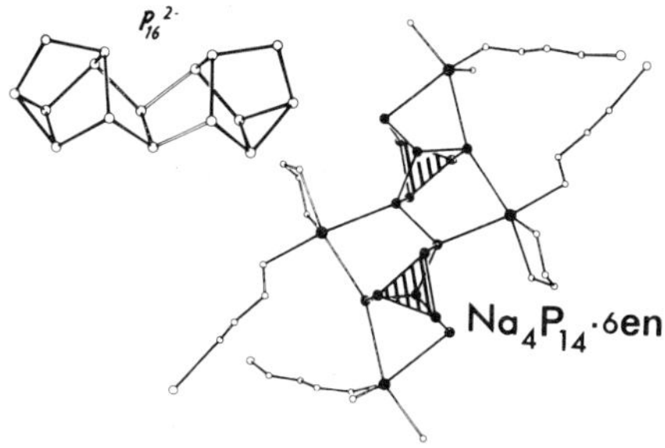

Figure 5. Structure of the P_{16}^{2-} anion (left) and structure of the solvated ion complex $Na_4P_{14} \cdot 6en$ (right), in which large black dots are Na atoms and shaded areas are P_7 units.

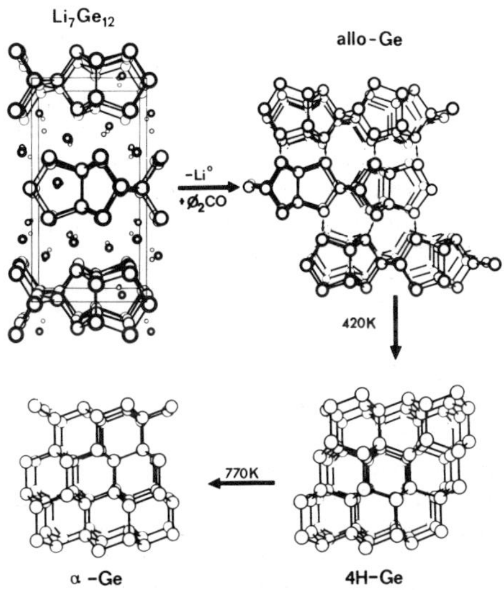

Figure 6. Formation of allo-Ge from Li_7Ge_{12} by a topochemical reaction and the main structural units of Li_7Ge_{12}, allo-Ge, 4H-Ge and α-Ge.

investigations we have therefore studied the metal phosphides M_3P_7 and M_3P_{11} which are distinguished by the appearance of the very fascinating cage systems P_7^{3-} and P_{11}^{3-} (Table I). An important result is the clear proof that the compounds M_2P_5 (M=alkaline metal), quoted in all text books, definitely do not exist. In all cases the previously described bright yellow compounds have the composition M_3P_7. The chemical analysis of previous works (9) agrees considerably better with the composition 3:7 than with composition 2:5. Obviously - and this is quite a general problem - the ratio 2:5 was considered as being simpler and more reasonable. The phosphides M_3P_{11} are usually orange coloured and are formed under very similar thermal conditions (21). The majority of the M_3P_7 and M_3P_{11} phosphides go through a crystalline-to-plastic first-order phase transition (17), (22).

The preparation of single crystals is difficult, but is successful in some case, so that we are well informed about the structures (1,6,17,21,22,23). The structures of the plastic phases are related to the well-known intermetallic phase Li_3Bi, where the centres of the polycyclic P_7^{3-} or P_{11}^{3-} anions surround the positions of the Bi atoms in Li_3Bi. The orientation of the polyanions is disordered (dynamically?). For these structures this orientation leads to a typical electron density distribution of a seemingly octahedral unit. In contrast the orientation of the anions is fixed for the crystalline phases. The symmetry of the unit cells as well as the distribution of cations and anions in these M_3P_7 and M_3P_{11} type structures reflect the direct relationship to the structures of the plastic phases.

The yellow phosphides M_3P_7 differ from all other phosphides by a remarkable property: they sublime congruently dissociative (cf. NH_4Cl). In the gas phase only M,g and P_4,g appear, whereas under similar conditions the phosphides M_3P_{11} dissociate into M_3P_7 and P_4 (24). These compounds can also be obtained as amorphous products by sublimation under suitable conditions. It has been shown that these amorphous products are extraordinarily reactive, therefore they are the most suitable starting materials for the transformations discussed above.

On the other hand however, the cluster-anions P_7^{3-} and P_{11}^{3-} are thermally remarkably stable. In the condensed state (in the crystal as well as in melts), the characteristic vibrations can be observed both in i.r. spectra and in Raman spectra upto temperatures of 900 K (25,26,27). As an example, the Raman spectra of Na_3P_7 in Figure 7 clearly show that the typical cluster-vibrations of the P_7^{3-}-anion are maintained up to the region of the plastic phase, although the absorption bands become increasingly broader and less distinct with temperature. The lattice vibrations at 50-100 cm^{-1} behave completely differently. As expected they disappear at the transition to the plastic phase. Completely unexpected however, they remain sharply resolved up to the critical temperature T_c. This effect can be connected with the presence of two undamped lattice modes (25).

Table I. Physical Properties of Crystalline and Plastic Phases of Alkaline Metal Compounds with Phosphorus and Arsenic

Compound	T [K]	color	state	structure symbol	space group	unit cell dimensions [Å] at T			
						a	b	c	T
Li_3P_7	< 900	yellow	crystal.	oP40	$P2_12_12_1$	9.750(3)	10.532(3)	7.596(3)	293 K
Li_3P_7	> 900	?	plastic.	?	?	?			
Na_3P_7	< 765	yellow	crystal.	oP80	$Pca2_1$	13.740(4)	10.391(3)	13.446(3)	293 K
Na_3P_7	> 765	black	plastic.	cF40	Fm3m	10.105(6)			873 K
Na_3P_{11}	< 853	orange	crystal.	oP56	Pnab	10.410(2)	12.466(3)	9.828(2)	293 K
Na_3P_{11}	> 853	black	plastic.	tP28	P - - -	10.21(1)		7.31(1)	873 K
K_3P_7	< 518	yellow	crystal.	?	?	?			
K_3P_7	> 518	dark yel.	plastic.	cF40	Fm3m	10.603(2)			540 K
K_3P_{11}	293	yellow	crystal.	oP56	Pnab	10.315(3)	13.940(5)	10.450(6)	293 K
Rb_3P_7	293	yellow	plastic.	cF40	Fm3m	10.818(2)			293 K
Rb_3P_{11}	293	orange	plastic.	cF56	Fm3m	11.592(2)			293 K
Cs_3P_7	293	yellow	plastic.	cF40	Fm3m	11.167(3)			293 K
Cs_3P_{11}	293	orange	plastic.	cF56	Fm3m	11.95(4)			293 K
Rb_3As_7	< 608	dark red	crystal.	?	?	?			
Rb_3As_7	> 608	black	plastic.	cF40	Fm3m	11.36(3)			673 K

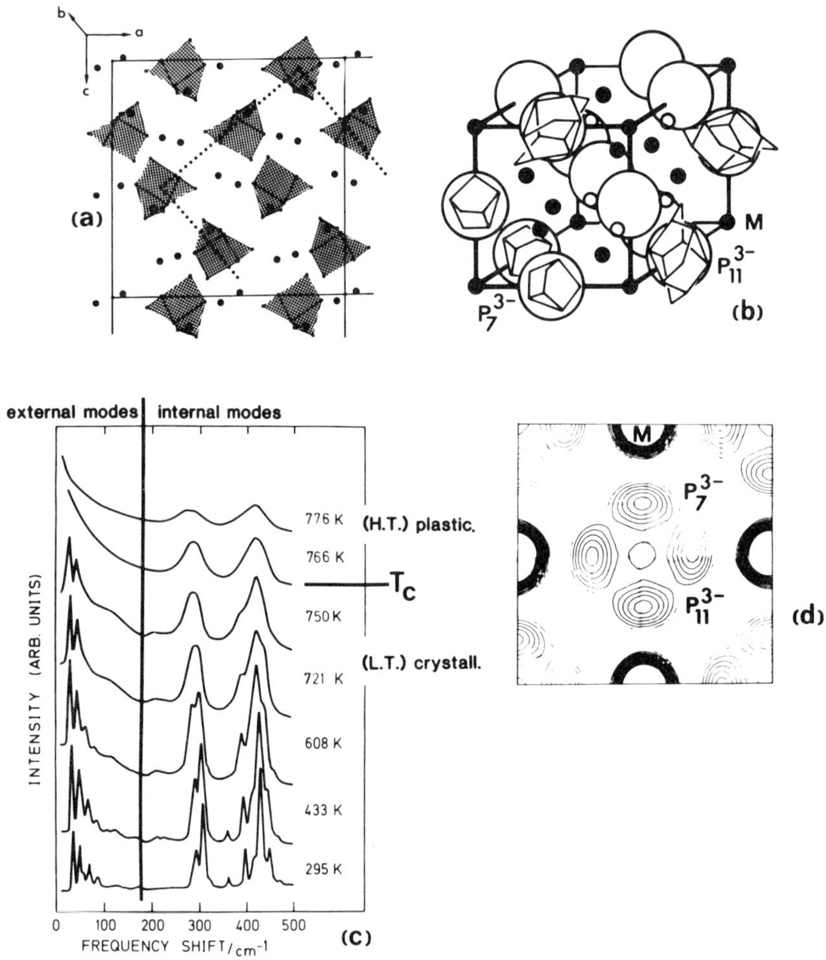

Figure 7. Crystalline (LT) and plastic (HT) phases of M_3P_7 and M_3P_{11}. Key: a, structure of the orthorhombic LT-Na_3P_7 as an example for the ordered crystalline low-temperature modifications (See *Table I. The relation of the LT-phase to the HT-phase is indicated by a dotted line.); b, structures of the cubic HT-M_3P_7 and HT-M_3P_{11} as examples for the plastically crystalline phases; c, temperature-dependent Raman spectra of Na_3P_7; and d, typical electron density in the plastically crystalline phases.*

Outlook

The application of solid compounds with cluster units as sources for new molecular compounds, as well as for new polymers opens up an attractive field in chemistry. In view of the wealth of binary and ternary solid compounds, there is no end in sight. The wide range of materials should encourage many chemical groups to work much more in this direction. Especially fascinating is the possibility to form new modifications of the elements which as metastable systems could also have interesting chemical and physical properties.

The phases of the 4B elements are largely unused, e.g. NaSi etc. with the tetrasilatetrahedrane Si_4^{4-} unit or $Li_{12}Si_7$ and $Li_{12}Ge_7$ with the trigonal planar Si_4 and Ge_4 stars (28). We were already able to observe reactions of these compounds, but upto now the products have not been characterized. Reactions of Na_8Si_{46}, Na_xSi_{136} (x=3-12) and Li_2Sn_5 with benzophenone proceed in the way described above, but so far only amorphous precipitations of silicon and tin have been obtained.

The search for suitable borides of alkali metals recently led to a new lithium boride with hexagonal B_6-octahedra framework unknown until now. Strong evidence for the existence of planar Li_3B molecules in the wide channels of the framework is given by the structural analysis (29). This compound points out another aspect involving reactions of solid phases; namely, the possible release of small molecules, which are clathrate-like, locked in the structure and are therefore stabilized.

Literature Cited

1. v.Schnering, H.G., Angew.Chem. 1981, 93, 44; Angew.Chem.Int. Ed.Engl. 1981, 20, 33
2. Fritz, G. and Hölderich, W., Naturwissenschaften 1975, 62, 573
3. v.Schnering, H.G., Fenske, D., Hönle, W., Binnewies, M., Peters, K., Angew.Chem. 1979, 91, 755; Angew.Chem.Int.Ed.Engl. 1979, 18, 679
4. Weber, D., Mujica, C., v.Schnering, H.G., Angew.Chem. 1982, 94 869; Angew.Chem.Suppl. 1982, 1801-1912; Angew.Chem.Int.Ed. Engl. 1982, 21, 863
5. v.Schnering, H.G., Hönle, W., Weber, D., Manriquez, V., Mujica, C. (to be published)
6. Manriquez, V., Ph.D., Thesis, University of Stuttgart, Stuttgart 1983
7. Mujica, C., Ph.D., Thesis, University of Stuttgart, Stuttgart 1983
8. Fritz, G., Uhlmann, R., Z.anorg.allg.Chem. 1980, 465, 59 Hönle, W., v.Schnering, H.G., Z.anorg.allg.Chem. 1980, 465, 72
9. Royen, P., Z.anorg.allg.Chem. 1938, 235, 324, footnote on page 331

10. Zintl, E., Goubeau, J., Dullenkopf, W., Z.Phys.Chem. 1931, A154, 1; Zintl, E., Harder, A., ibid. 1931, A154, 47; Zintl, E., Dullenkopf, W., Z.Phys.Chem. 1932, B16, 183; Zintl, E., Kaiser, H., Z.anorg.allg.Chem. 1933, 211, 113
11. Corbett, J.D., Adolphon, D.G., Merryman, D.J., Edwards, P.A., Armatis, F.J., J.Am.Chem.Soc., 1975, 97, 6267
12. Diehl, L., Khodadedeh, K., Kummer, D., Strähle, J., Chem.Ber. 1976, 109, 3404
13. Baudler, M., Pontsen, T., Hahn, J., Ternberger, W., Faber, F., Z.Naturforschung 1980, B35, 517
14. Baudler, M., Angew.Chem. 1982, 94, 520
15. Böhm, M.C., Gleiter, R., Z.Naturforschung 1981, B36, 498
16. Hönle, W., v.Schnering, H.G. (to be published)
17. v.Schnering, H.G., Hönle, W., Manriquez, V., Meyer, T., Mensing, Ch., Giering, W., Coll.Abstracts II. European Conf. Sol.State Chem., Eindhoven 1982
18. v.Schnering, H.G., Manriquez, V., Hönle, W., Angew.Chem. 1981, 93, 606; Angew.Chem.Int.Ed.Engl. 1981, 20, 594
19. Grüttner, A., Nesper, R., v.Schnering, H.G., Angew.Chem. 1982, 94, 933; Angew.Chem.Int.Ed.Engl. 1982, 21, 912
20. Grüttner, A., Ph.D., Thesis, University of Stuttgart, Stuttgart 1982
21. v.Schnering, H.G., Chapter 14 in A.L. Rheingold (Ed.) Homoatomic Rings, Chains and Macromolecules of Main Group Elements, Elsevier, Amsterdam 1977
22. Meyer, T., Ph.D., Thesis, University of Stuttgart, Stuttgart 1983
23. Wichelhaus, W., v.Schnering, H.G., Naturwissenschaften 1973, 60, 104
24. Santandrea, R., v.Schnering, H.G. (to be published)
25. Henkel, W., Strößner, K., Hochheimer, H.D., Hönle, W., v.Schnering, H.G., Proc. of the 8th Int.Conf.Raman Spectroscopy, Bordeaux 1982; Eds. J. Lascombe, P. Vhuong, J. Wiley, New York 1982
26. Bues, W., Somer, M.S., Brockner, W., Grünewald, D. Naturwissenschaften 1977, 64, 583
27. Somer, M.S., Ph.D., Thesis, Technical University Clausthal, Clausthal 1979
28. v.Schnering, H.G., Nesper, R., Curda, J., Tebbe, K.F., Angew.Chem. 1980, 92, 1070; Angew.Chem.Int.Ed.Engl. 1980, 19, 1033
29. v.Schnering, H.G., Mair, G., (to be published)

RECEIVED April 27, 1983

5

Cyclophosphathiazenes Containing Two- or Three-Coordinate Sulfur

N. BURFORD, T. CHIVERS, M. N. S. RAO, and J. F. RICHARDSON

University of Calgary, Department of Chemistry, Calgary, Alberta T2N 1N4 Canada

The preparation, molecular and electronic structures, and some reactions of cyclophosphathiazenes containing two or three coordinate sulfur are described. Inorganic heterocycles of general formulae $(R_2PN)_n(SN)_2$ or $(R_2PN)_x(NSCl)_y$ are prepared by the reactions of S_4N_4 with phosphorus(III) reagents [e.g. R_2PPR_2(R=Me,Ph) or Ph_2PH, n=1,2; or Ph_2PCl, x=2, y=1]. In addition to substitution at sulfur, $(Ph_2PN)_2NSCl$ readily undergoes ring opening reactions to give novel bicyclic, tricyclic, or spirocyclic PSN ring systems. X-ray crystallographic data and the results of MO calculations show that the excess of electrons may be accommodated by partial occupation of π^*-levels or by formation of transannular S-S bonds in these π-electron rich heterocycles.

A wide variety of cyclic S-N compounds containing two or three coordinate sulfur is known (1). The binary S-N rings vary in size from four (S_2N_2) to ten ($S_5N_5^+$) atoms and examples of anions ($S_3N_3^-$) and cations ($S_3N_2^+$, $S_4N_3^+$, $S_4N_4^{2+}$) are known in addition to neutral molecules (2). The excess of electrons in these planar, electron-rich heterocycles is often accommodated in low lying π^* orbitals but, in some cases (e.g. S_4N_4, $S_4N_5^-$), the number of π^* electrons is reduced by formation of transannular S-S bonds to give cages (2,3).

An even broader range of ring sizes is found for the cyclophosphazenes $(R_2PN)_x$ (x=3-17) but these inorganic heterocycles are π-electron precise (i.e. the number of π-electrons is equal to the number of atomic centers in the ring system). Although both bicyclic and condensed structures have been identified, simple anions or cations of cyclophosphazenes are unknown (4).

The combination of the monomer units R_2PN and NS provides many possibilities for inorganic ring systems containing phosphorus, nitrogen, and two coordinate sulfur. Thus one might an-

0097-6156/83/0232-0081 $6.00/0.

ticipate a homologous series $(R_2PN)_n(SN)_2$ in which the first member is a six-membered ring (1, n=1) and the second member is an eight-membered ring with structural isomers (2 and 3, n=2). In addition, larger ring sizes (n=3,4 etc.) are possible.

Similarly the combination of R_2PN and NSCl units affords heterocycles containing phosphorus, nitrogen and three coordinate sulfur (e.g. 4 and 5) which may be considered as hybrids of the well known ring systems $(R_2PN)_3$ and $(NSCl)_3$.

This paper describes the preparation, molecular and electronic structures and some reactions of compounds 1 - 5, for which the generic name cyclophosphathiazenes will be used.

Cyclophosphathiazenes with Two Coordinate Sulfur

Preparation. The first report of a heterocycle of this type (1, R = Me_3SiNH) involved the reaction of S_4N_4 with the phosphorus(V) reagent $(Me_3Si)P(NSiMe_3)_2$ (5). The structure of this deep blue compound consists of a highly puckered six-membered ring (6). The reaction of R_2PPR_2 with S_4N_4 provides a more general route to such heterocycles. Thus treatment of S_4N_4 with tetraphenyl- or tetramethyl-diphosphine in toluene at reflux produces 1 (R=Ph,Me) in modest yields (15 - 20%, 7,8). The Ph_2PPPh_2/S_4N_4 reaction mixture was monitored by ^{31}P nmr spectroscopy which revealed the presence of $(Ph_2PN)_3$, $(Ph_2PN)_4$, and $(Ph_2PS)_2$ in addition to several other products (9). Two of these products have been separated by gel permeation chromatography and identified as the structural isomers 2 and 3 (R=Ph,Me). The reaction of Ph_2PH with S_4N_4 also produces 1 - 3 and is preferred for the preparation of 1 and the 1,3-isomer, 2, because the products are more readily separated from $Ph_2P(S)H$ than from $(Ph_2PS)_2$. However, the Ph_2PH/S_4N_4 reaction gives a very low yield of the 1,5-isomer, 3.

Molecular Structures. An X-ray structural determination of 1(R=Ph) shows that the NSNSN sequence of the six-membered ring is essentially planar but the phosphorus atom is tilted 0.28Å out of this plane (8). The P-N and S-N bond distances are ca. 1.62 and 1.57Å, respectively, compared to ca. 1.60Å for the P-N distance in $(Ph_2PN)_3$ (10) and ca. 1.60Å for the S-N bond length in $S_3N_3^-$ (11). These trends in bond lengths are consistent with expectations based on the view that $(Ph_2PN)_3$, $Ph_2PS_2N_3$ and $S_3N_3^-$ are 6, 8, and 10 π-electron systems, respectively (7,8).

The structure of the orange 1,3-isomer of the eight-membered ring, 2, resembles that of the six-membered ring in that the NSNSN unit is essentially planar. The phosphorus atoms lie out and on opposite sides of this plane by ca. 0.70Å (9). The structural parameters for 2, are given in Figure 1.

In contrast, the structure of the pale yellow 1,5-isomer, 3, consists of a folded eight-membered ring with a transannular S-S bond (d(S-S)=2.53Å, Figure 2). The two planar S_2N_2 units intersect at an angle of ca. 116^0 and the phosphorus atoms lie ca.

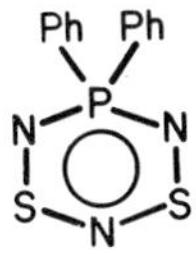

1

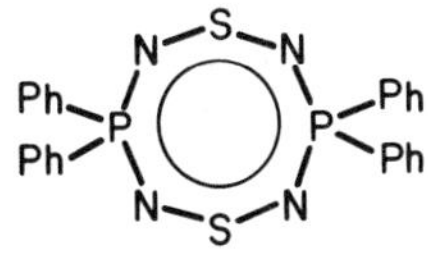

2

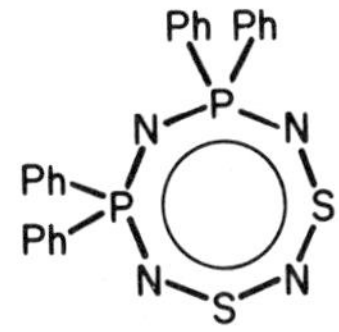

3

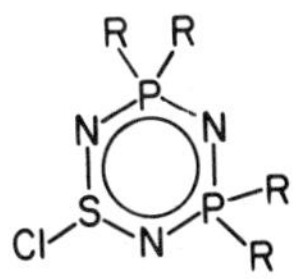

5

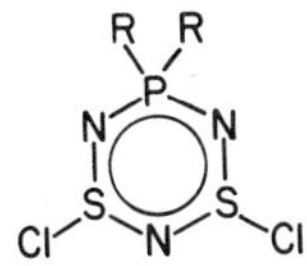

4

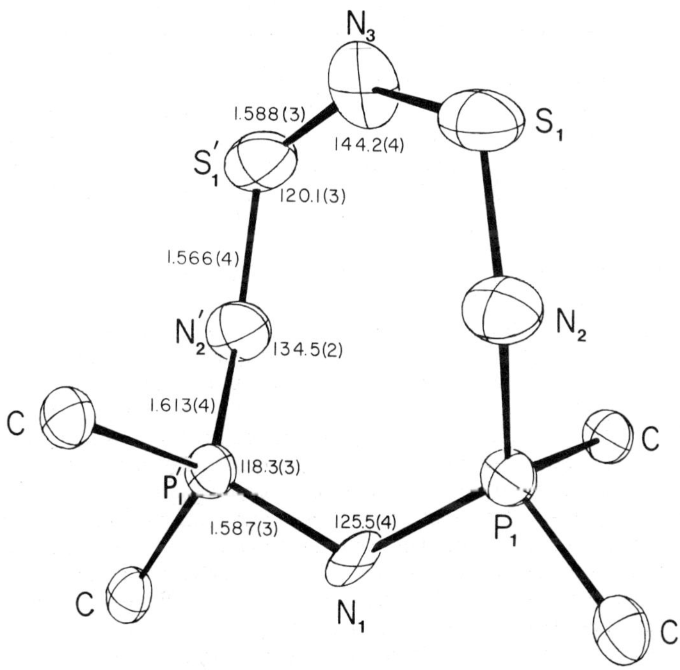

Figure 1. ORTEP drawing of 1,3-(Ph_2PNPPh_2)S_2N_3 showing bond lengths (Å) and angles (deg). All ORTEP figures are drawn with thermal ellipsoids at the 50% probability level, and only α-C atoms of the phenyl rings are shown.

0.21Å below the respective planes (9). The corresponding methyl derivative, 3 (R=Me), has a very similar structure except that the phosphorus atoms lie above and below the planes by ca. +0.19 and -0.48Å, respectively, apparently due to a solid state effect (12).

This structure may be compared with those of the recently reported isoelectronic carbon analogues $RC(NSN)_2CR$(13, R=Me_2N,Ph). In the carbon compounds the effect of the exocyclic substituent on ring conformation is remarkable. Thus the phenyl derivative is completely planar and has structural parameters consistent with a 10 π-electron system while the dimethylamino derivative has a folded structure with a transannular S-S distance of 2.43Å (Figure 3).

Electronic Structures. Ab initio HFS calculations for 1 (R = H) have confirmed the conclusions of an HMO analysis that the six-membered ring is an 8 π-electron system in which the HOMO is a π* orbital (8). These calculations provide the following additional insights into the nature of the bonding in 1 (8).

(i) The non-planar (experimental) conformation of the PS_2N_3 ring is slightly lower in energy (ca. 15 kcal mol^{-1}) than the planar ring, but the order of energy levels does not change with geometry.

(ii) The availability of 3d orbitals on phosphorus leads to a stabilization of the π-system in 1 compared to $S_3N_3^-$, especially the π* levels.

(iii) The strong visible absorption band observed at 545-580 nm for 1 (R = Me, Ph, PhO, Me_3SiNH) is assigned to the HOMO(π*)→LUMO(π*) transition (calculated value ca. 560 nm) on the basis of calculated transition moments.

(iv) Most of the π-electron density in 1 is located on the S_2N_3 unit suggesting that the internal salt $R_2P^+S_2N_3^-$ is a reasonable model for the more qualitative Hückel approach.

Justified by the similarity of the molecular structure of 2 with that of 1, the former can be considered to consist of an $NSNSN^-$ unit accommodating 8π-electrons interacting with the 2 π-electron unit $Ph_2P{=}N^+{=}PPh_2$. This interaction leads to a stabilization of the HOMO ($2a_2$) and LUMO ($4b_1$) levels in a manner similar to the lowering of the π-energy levels in 1 which results from the interaction of the 3d orbitals on phosphorus with the π-type orbitals of the $NSNSN^-$ unit (Figure 4). Thus the eight-membered ring can be described as a 10 π-electron system and, by analogy with 1, it seems reasonable to suggest that a similar π*→π* transition is responsible for the strong visible absorption band at ca. 460 nm and, hence, the orange color observed for the 1,3-isomer.

The existence of different conformations for the 1,5-isomer of the eight-membered rings can be understood in terms of HMO

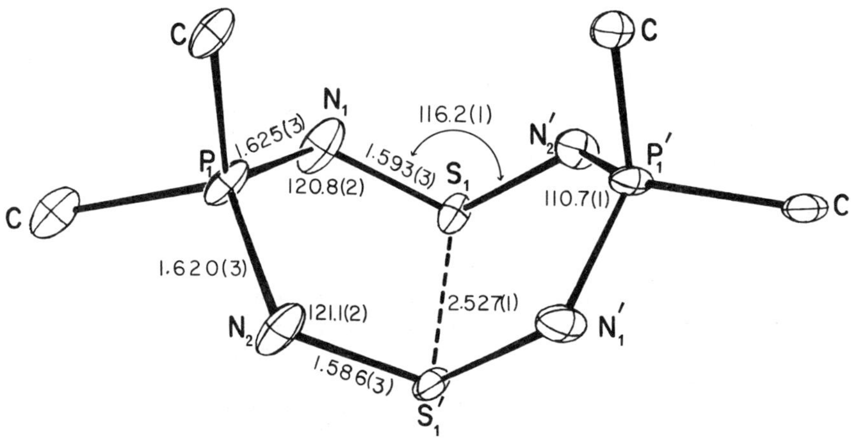

Figure 2. ORTEP drawing of 1,5-$Ph_2P(NSN)_2PPh_2$ showing bond lengths (Å) and angles (deg).

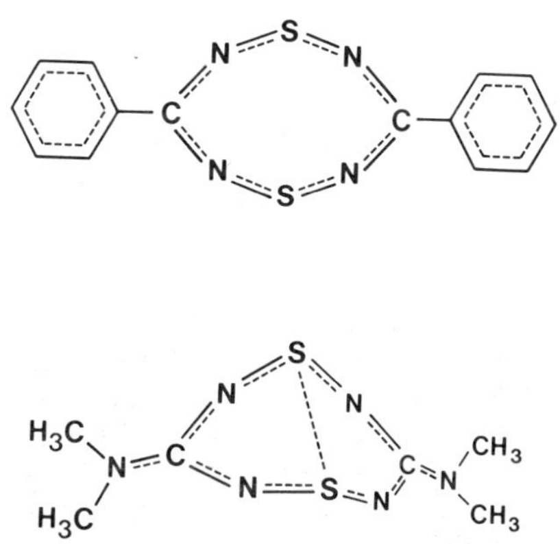

Figure 3. Structures of 1,5-$PhC(NSN)_2CPh$ (top) and 1,5-$Me_2NC(NSN)_2CNMe_2$ (bottom) (13).

calculations for the two limiting structures (a) a planar ten π-electron system and (b) a folded structure consisting of two independent four π-electron three center units plus an S-S σ-bond. Figure 5 shows a correlation diagram for these two structures. As illustrated by Gleiter for S_4N_4 (14), the HOMO (an anti-bonding π-level) of the planar molecule transforms into the bonding S-S σ-interaction of the folded ring. Consequently, the butterfly structure will be adopted if this σ-interaction is sufficiently strong to overcome the greater π-delocalisation energy of the planar relative to the folded ring (assuming no significant changes in the rest of the σ framework). The HOMO of the planar ring is the orbital most sensitive to changes in the electronegativity of E and it is stabilized by more electronegative values of α_E. Thus it is not unreasonable to find that the electronegative substituent Ph favors the planar structure in $RC(NSN)_2CR$, while the electron-releasing Me_2N group produces a folded conformation. The butterfly structure of 3 is probably a consequence of the lower electronegativity of phosphorus compared to carbon (12).

Cyclophosphathiazenes with Three Coordinate Sulfur

Preparation. Only one example of a cyclophosphathiazene of the type $(R_2PN)_x(NSCl)_y$ has been reported. The compound $(Cl_2PN)_2NSCl$ (5, R=Cl) was obtained as a colorless oil by the reaction of PCl_5 with $Me_3SiNSNSiMe_3$ (15). The related cation $(Cl_2PN)_2NS^+$, formed by treatment of this oil with $SbCl_5$, was shown to be a planar six-membered ring by X-ray crystallography (15).

An alternative route to heterocycles of type 5 is the reaction of S_4N_4 with trivalent phosphorus compounds containing P-Cl bonds (16). For example, the reaction of Ph_2PCl with S_4N_4 in a 3:1 molar ratio in CH_3CN for 3h at reflux produces an 85% yield of $(Ph_2PN)_2NSCl$ as moisture-sensitive, pale yellow crystals. The linear product $[Ph_2PClNPClPh_2]^+ Cl^-$ is also obtained and the yield of this material increases significantly as the molar ratio Ph_2PCl/S_4N_4 is increased.

Another member of the series (4, x=1, y=2) can be obtained by the smooth oxidative addition of Cl_2 to 1 using $PhICl_2$ (17).

The dichloro compound 4 can be converted to the bicyclic molecule 6 on treatment with $Me_3SiNSNSiMe_3$ (Figure 6). Related bicyclic compounds with fluorine (or Ph and F) substituents on phosphorus have previously been obtained from the reaction of PF_5 (or $PhPF_4$) with $Me_3SiNSNSiMe_3$ (18, 19). The thermal decomposition of 6 in toluene at ca. 90°C for 3h regenerates the six-membered ring (1).

Structure of $(Ph_2PN)_2NSCl$. An X-ray structural determination of $(Ph_2PN)_2NSCl$ shows it to consist of a six-membered ring in which the five atom NPNPN unit is planar to within 0.05Å while

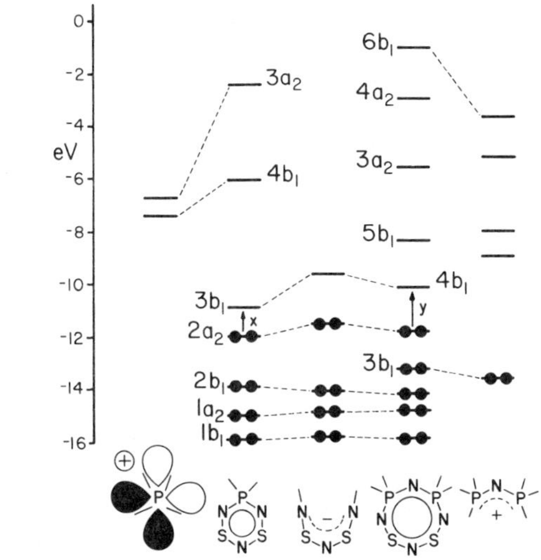

Figure 4. Orbital symmetry correlation diagram for $R_2PS_2N_3$ and 1,3-$(R_2PNPR_2)S_2N_3$.

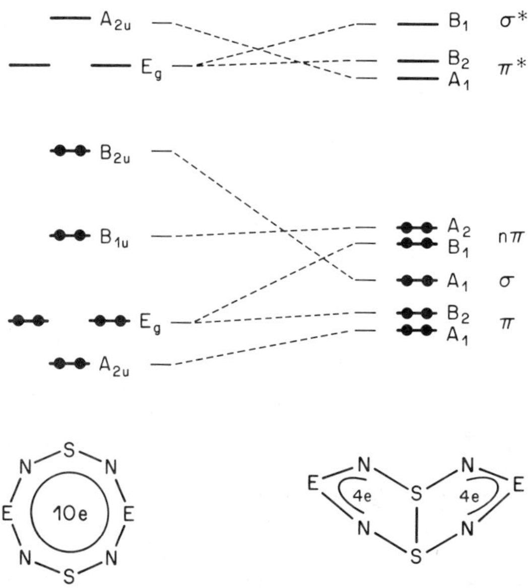

Figure 5. Orbital symmetry correlation diagram for the planar (left) and folded (right) models of $E(NSN)_2E$, where $\alpha_E = \alpha_S$.

(*Reproduced from* Inorg. Chem. *1982*, 21, *982*.)

the sulfur atom lies ca. 0.31Å out of this plane (Figure 7). The S-N bond distances (ca. 1.56Å) are somewhat shorter than the corresponding distances in $(NSCl)_3$ (1.60-1.61Å) (20), but the mean P-N distance in the PNP unit (ca. 1.59Å) is typical of that found for cyclophosphazenes (4). However, the P-N bonds linking this unit to the NSN unit are considerably longer (1.66-1.67Å) suggesting a tendency towards localization of π-bonding at opposite ends of the molecule. A similar difference in S-N and P-N bond lengths in P,N,S rings containing four coordinate sulfur has been attributed to the different electronegativities of the phosphorus and sulfur centers (21). The S-Cl distance (2.36Å) is substantially longer than the corresponding distance in $(NSCl)_3$ (2.08Å). The corresponding iodide $(Ph_2PN)_2NSI$ has very similar endocyclic bond lengths and an S-I distance of ca. 2.72Å (22).

Reactions of $(Ph_2PN)_2NSCl$. The monofunctionality of **5** (R=Ph) provides a unique opportunity to investigate reactions at the sulfur center of the P_2SN_3 ring and preliminary results indicates that both direct substitution and ring opening reactions can occur.

For example, the reaction of **5** with KI in CH_3CN gives an almost quantitative yield of dark red crystals of the corresponding iodide, $(Ph_2PN)_2NSI$, while a very moisture-sensitive azide can be obtained using Me_3SiN_3 (22).

The thermolysis of $(Ph_2PN)_2NSCl$ above 200°C, or of the corresponding iodide at lower temperatures (ca. 140°C), produces colorless crystals of a very stable compound which has been shown by X-ray crystallography to have a spirocyclic structure in which two six-membered rings are joined at a common sulfur atom (Figure 8). This compound is frequently found as a product of transformations involving cyclophosphathiazenes. Each six-membered ring is formally associated with 6 π-electrons, i.e. π-electron precise and isoelectronic with $(R_2PN)_3$, and this may account for the high stability of the spirocycle.

Another interesting example of a ring opening reaction occurs when $(Ph_2PN)_2NSCl$ is treated with $Me_3SiNSNSiMe_3$. A tricyclic compound, whose structure is illustrated in Figure 9, is formed in ca. 20% yield. This molecule consists of two eight-membered rings (1,3-isomer, **2**) fused at a common sulfur atom and linked by an S-S interaction (S(2)-S(3) = 2.30Å) to give a central five-membered S_3N_2 ring. This compound can be formally derived from the six-membered spirocycle by insertion of an NS unit into each of the six-membered rings. Thus, each eight-membered ring would be formally associated with 9 π-electrons (i.e. π-electron rich). The formation of the inter-ring S-S bond probably has the effect of converting two weakly anti-bonding (π*) electrons to an S-S σ-bonding electron pair (cf. formation of transannular S-S bonds in **3**).

In the attempted coupling of two six-membered P_2SN_3 rings through an S-S bond by treatment of $(Ph_2PN)_2NSCl$ with Ph_3Sb an

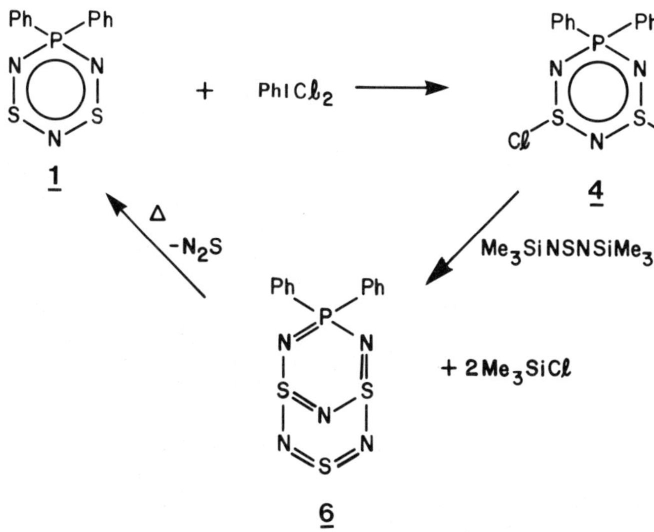

Figure 6. Conversion of $Ph_2PS_2N_3$ to the bicyclic molecule $Ph_2PS_3N_5$ by using the reagents $PhICl_2$ and $Me_3SiNSNSiMe_3$.

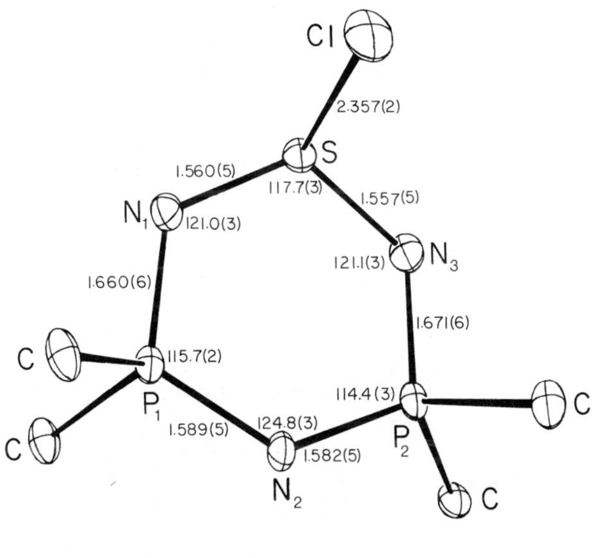

Figure 7. ORTEP drawing of $(Ph_2PN)_2NSCl$ showing bond lengths (Å) and angles (deg).

(Reproduced with permission from J. Chem. Soc., Chem. Commun. *1982, 982.)*

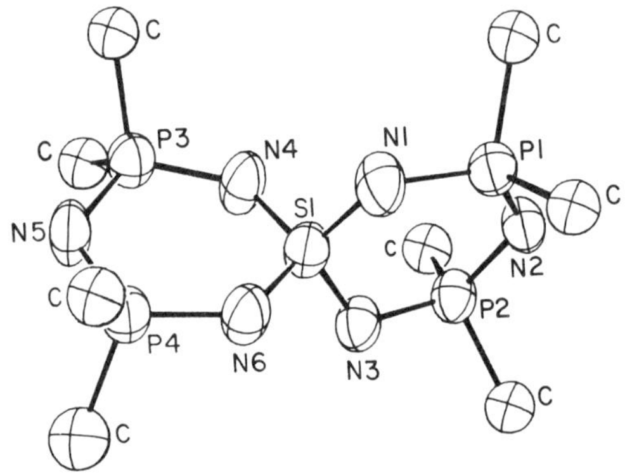

Figure 8. ORTEP drawing of spirocyclic compound formed by thermal decomposition of $(Ph_2PN)_2NSX$, where X = Cl or I.

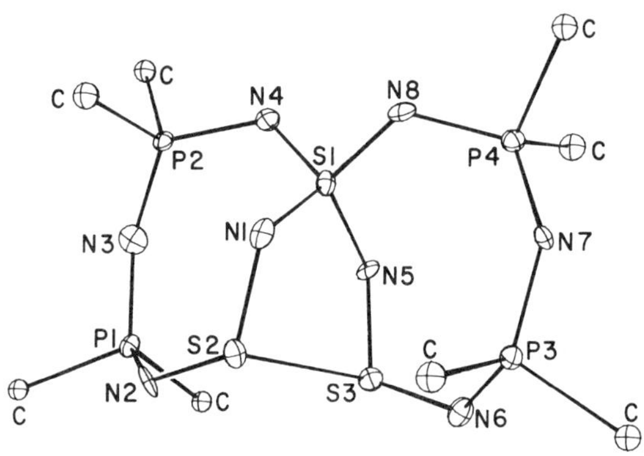

Figure 9. ORTEP drawing of tricyclic compound obtained from $(Ph_2PN)_2NSCl$ and $Me_3SiNSNSiMe_3$ (16).

isomer of the expected product was obtained (Figure 10). This compound is a twelve-membered ring with a transannular S-S bond of 2.39Å cf. d(S-S) = 2.53Å in 3. It can be viewed as resulting from the insertion of an additional Ph_2PN unit into each half of 3. This ring expansion causes the molecule to twist about the S-S bond. However, the mean S-N bond length of 1.598(2)Å is similar to that found for 3 (1.590(3)Å) and the mean P-N bond lengths of the NPNPN unit [1.582(3) (internal) and 1.619(3)Å (terminal)] are indistinguishable from those found for 2 [1.587(3) and 1.613 (3)Å, respectively]. Since a planar, monocyclic $P_4N_2S_6$ ring would have 14 π-electrons (cf. $S_5N_5^+$, 23), it is tempting to suggest that the formation of a transannular S-S bond provides a larger stabilization than the greater π-delocalisation energy of the planar twelve-membered ring (cf. 3). It seems likely, however that other factors also contribute to the observed geometry.

Finally, it should be pointed out that this twelve-membered ring belongs to the homologous series $(R_2PN)_n(SN)_2$ (n=4) and thus the prediction that larger ring sizes may be anticipated in this series has been vindicated.

Summary and Conclusions

The first members of a potentially extensive homologous series of cyclophosphathiazenes containing two coordinate sulfur, $(Ph_2PN)_n(SN)_2$ (n=1,2,4) have been prepared and structurally characterized. These heterocycles are π-electron rich, i.e. the

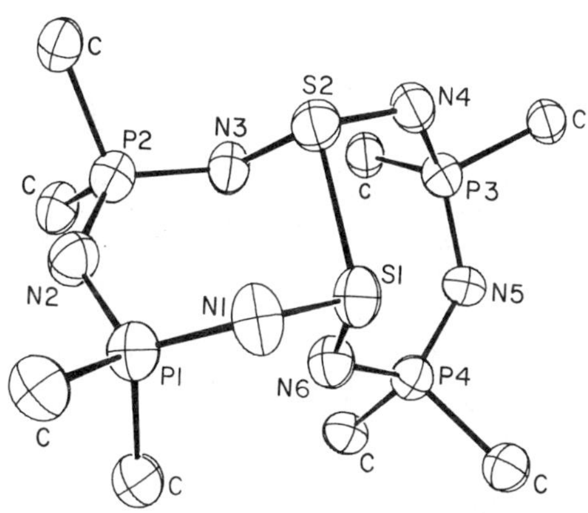

Figure 10. ORTEP drawing of 12-membered ring obtained from $(Ph_2PN)_2NSCl$ and Ph_3Sb.

number of π-electrons exceeds the number of atomic centers in the ring system, and the excess electrons may be accommodated either by partial occupation of π*-levels or, in favourable cases, by the formation of transannular S-S bonds.

Cyclophosphathiazenes containing three coordinate sulfur, which are hybrids of $(Ph_2PN)_3$ and $(NSCl)_3$, have also been prepared. The inorganic heterocycle $(Ph_2PN)_2NSCl$ readily undergoes ring opening to give the π-electron precise spirocycle $(Ph_2PN)_4SN_2$ or bicyclic and tricyclic molecules in which inter- or intra-ring S-S σ-bonds are a dominant structural feature.

Literature Cited

1. Roesky, H.W. Adv. Inorg. Chem. Radiochem. 1979, 22, 239.
2. Gleiter, R. Angew. Chem., Int. Ed. Engl. 1981, 20, 444.
3. Chivers, T.; Oakley, R.T. Top. Curr. Chem. 1982, 102, 117.
4. Krishnamurthy, S.S.; Sau, A.C.; Woods, M. Adv. Inorg. Chem. Radiochem. 1978, 21, 41.
5. Appel, R.; Halstenberg, M. Angew. Chem., Int. Ed. Engl. 1976, 15, 696.
6. Weiss, J. Acta Crystallogr., Sect. B 1977, B33, 2272.
7. Burford, N; Chivers, T.; Oakley, R.T.; Cordes, A.W.; Swepston, P.N. J. Chem. Soc., Chem. Commun. 1980, 1204.
8. Burford, N.; Chivers, T.; Cordes, A.W.; Laidlaw, W.G.; Noble, M.C.; Oakley, R.T.; Swepston, P.M. J. Am. Chem. Soc. 1982, 104, 1282.
9. Burford, N.; Chivers, T.; Richardson, J.F. Inorg. Chem. 1983, 22, in press.
10. Ahmed, F.R.; Singh, P.; Barnes, W.H. Acta Crystallogr., Sect. B 1969, B25, 316.
11. Bojes, J.; Chivers, T.; Laidlaw, W.G.; Trsic, M. J. Am. Chem. Soc. 1979, 101, 4517.
12. Burford, N.; Chivers, T.; Codding, P.W.; Oakley, R.T. Inorg. Chem. 1982, 21, 982.
13. Ernest, I,; Holick, W.; Rihs, G.; Schomburg, G.; Shokam, G.; Wenkert, D.; Woodward, R.B. J. Am. Chem. Soc. 1981, 103, 1540.
14. Gleiter, R. J. Chem. Soc. A 1970, 3174.
15. Pohl, S.; Peterson, O.; Roesky, H.W. Chem. Ber. 1979, 112, 1545.
16. Chivers, T.; Rao, M.N.S.; Richardson, J.F. J. Chem. Soc., Chem. Commun. 1982, 982.
17. Oakley, R.T. personal communication.
18. Appel, R.; Ruppert, I.; Milker, R.; Bastian, V. Chem. Ber. 1974, 107, 380.
19. Weiss, J.; Ruppert, I.; Appel, R. Z. Anorg. Allg. Chem. 1974, 406, 329.

20. Wiegers, G.A.; Vos, A. Acta Crystallogr., Sect. B 1966, B20, 192.
21. Van de Grampel, J.C. Rev. Inorg. Chem. 1981, 3, 1.
22. Chivers, T.; Rao, M.N.S.; Richardson, J.F. umpublished results.
23. Bartetzko, R.; Gleiter, R. Inorg. Chem. 1978, 17, 995.

RECEIVED April 15, 1983

6

Polyatomic Zintl Anions Stabilized Through Crypt Complexation of the Cation

Heteroatomic Examples

JOHN D. CORBETT, SUSAN C. CRITCHLOW, and ROBERT C. BURNS[1]

Ames Laboratory–Department of Energy and Iowa State University, Department of Chemistry, Ames, IA 50011

A summary is given of the structures for the known homopolyatomic anions of the representative elements from groups IV and V and for a variety of heteropolyatomic examples. Also considered are factors important in their stabilization, electronic requirements, isoelectronic analogs, and new results for Sb_7^{3-}, Sb_4^{2-}, $Pb_2Sb_2^{2-}$ and the unusual $^1_\infty[KSn_9^{3-}]$. The contributions of Ralph Rudolph to the study of these anions in solution are noted.

The first evidence that post-transition elements, the metals especially, could be reduced to highly colored anions was published over 90 years ago by Joannis (1) who discovered that sodium and lead or their alloys dissolve in liquid ammonia to yield an intensely green-colored solute. A stoichiometry of 2.25 lead atoms per sodium (2) for what was evidently an anion led Kraus (3) to formulate this as Pb_9^{4-}. Until the past decade the principal information regarding this and many other species were the stoichiometries obtained by Zintl and coworkers (4,5,6) from both potentiometric titrations of the ammonia solutions and exhaustive extraction of alloys followed by analysis. Among those so identified were Sn_9^{4-}, $Pb_{7,9}^{4-}$, $Sb_{5,7}^{3-}$ and $Bi_{3,5,7}^{3-}$. But they found that structural studies were precluded by the characteristic decomposition of the amorphous solids isolated from solution into intermetallic phases and, at times, the heavy element, e.g.,

$$(Na(NH_3)_n^+)_3Sb_7^{3-} \longrightarrow NaSb(s) + 2NaSb_3(s) + 3nNH_3(g)$$

[1]Current address: McMaster University, Department of Chemistry, Hamilton, Ontario L8S 4M1, Canada

0097–6156/83/0232–0095 $06.00/0

The isolation of a few interesting and presumably related compounds from similar ethylenediamine (en) solutions has been described more recently (7,8), e.g., $Na_4Sn_9 \cdot 7en$ and $Na_3Sb_7 \cdot 4en$ plus a $Na_4Ge_9 \cdot 5en$ of marginal stability. An incomplete structural study of the first revealed a distorted Sn_9 polyhedron with large thermal ellipsoids for two tin atoms, some residual Na-Sn interactions, and substantial disorder of the en which apparently complexed the sodium.

It was the recognition (9) that 2,2,2-crypt (cp,I) would be

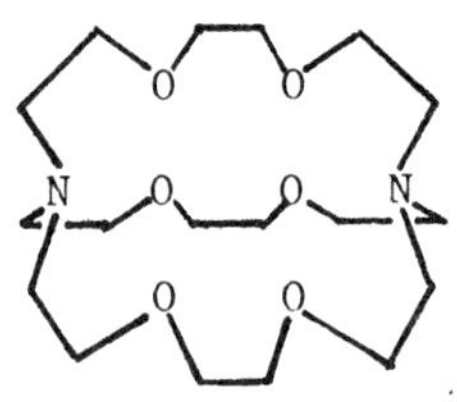

I

a much more effective way to sequester the sodium or potassium cations and to prevent reversion to conventional phases that has led to the isolation of the large number of stable compounds of the polyanions which are to be considered herein. Typical synthesis reactions may be relatively fast and simple, e.g.,

$$NaSn_{2.25}(\text{alloy}) + 2,2,2\text{-crypt} \xrightarrow{en} \xrightarrow{evap} (cp\text{-}Na^+)_4Sn_9^{4-}$$

Homopolyatomic Anions

The homopolyatomic or Zintl anions incorporating post-transitions elements of groups IV and V which have been so synthesized and structurally characterized are listed in Table I, together with their observed symmetries and other ions or molecules with which they are isoelectronic. As a matter of fact only three of those given, Sn_9^{4-}, Sb_7^{3-} and Bi_4^{2-}, were observed in the earlier solutions studies (6), the last corresponding to an analysis of 1.95 Bi atoms per charge which they attributed to "Bi_7^{3-}" (15). All of the ions listed appear to require crypt or equivalent complexation of the alkali metal cations to prevent the afore-mentioned decomposition. Although replacement of the alkali metal cation by a nonmetallic one would presumably obviate the requirement of a well-complexed metal cation, the compound $(NMe_4)_4Sn_9$ was recently found to decompose slowly at room temperature to various methyl stannanes (18), while the introduction of ϕ_4P^+ or ϕ_4As^+ to Sn_9^{4-} solutions in en also causes decomposition (8). Of course, many polyphosphides and at least some polyarsenides are known to be stable without sequestering the cation (19) and the same appears to be generally true of

Table I
Homopolyatomic Anions of Known Structure From Groups IV and V[a]

Anion	Symmetry	Characteristics	Isoelectronic Analogs	Str. Ref.
Ge_9^{2-}	$C_{2v}(\sim D_{3h})$	dark red rods	$B_9H_9^{2-}$	10
Ge_9^{4-}	C_{4v}		$Bi_9^{5+}(D_{3h})$	
Ge_9^{2-}	$\sim D_{3h}$[b]	lavender chunks	$B_9H_9^{2-}$	11
Sn_9^{4-}	C_{4v}	dark red rods	$Bi_9^{5+}(D_{3h})$	12
Sn_5^{2-}	D_{3h}	orange-brown plates	Bi_5^{3+}	13
Pb_5^{2-}	D_{3h}	ruby red prisms		13
Sb_4^{2-}	D_{4h}	dark red plates	Se_4^{2+}, Te_4^{2+}	14
Bi_4^{2-}	D_{4h}	dark green prisms or plates		15
As_{11}^{3-}	$\sim D_3$	deep red rods	P_{11}^{3-}	16
Sb_7^{3-}	C_{3v}	brown prisms, rods	P_7^{3-}, P_4S_3	14, 17

[a]Crypt-K$^+$ or crypt-Na$^+$ salts
[b]Partially disordered

polychalcogenide anions, so both groups are excluded from the present consideration.

The occurrence of many isoelectronic and often isostructural analogs of these anions in Table I is reassuring. These are found principally among homopolyatomic cations and electronically related polyboranes, for which the interrelationships have been previously discussed (20). The contrast in experimental conditions necessary to stabilize the two types of polyatomic ions is worth emphasizing. The nature of the counterion and the solvent, if any, are critical to both. The cations require high acidity - the absence of basic molecules or anions - and can generally be obtained from acidic halide and other melts containing mainly $AlCl_4^-$, $Al_2Cl_7^-$, $HfCl_6^{4-}$, etc. anions (superacids appear useful in an analogous way mainly for the polychalcogenide cations). On the other hand, synthesis of the polyanions needs just the opposite, a good polar solvent and a strongly complexing ligand for the countercation in order to block electron transfer from the anion and thereby to destabilize what are evidently strong interactions and delocalization encountered in the dense solid state. In no instance has coordination of en (or NH_3) to a polyanion been observed in spite of the fact that each atom in these ions has only three to five nearest neighbors. According to earlier conclusions (summarized in 20), this must result from the presence of a small orbital basis set - substantially p orbitals only - which is important to the stability of these polyatomic species, anions as well as cations. On the other hand a basic functionality is probably present in some, for example at the open square base of Sn_9^{4-}.

There are no particular surprises regarding the geometry of any of the Zintl ions tabulated although some of these had previously been represented only by compounds of nonmetals. The anions in $(cp\text{-}Na^+)_4Sn_9^{4-}$ and a recent result for $(cp\text{-}K^+)_3Sb_7^{3-}\cdot 2en$ are shown in Figure 1. The configurations of all of the anions are in accord with Wade's rules, Ge_9^{4-} and Sn_9^{4-} representing the first examples of C_{4v} geometry predicted for a 22-electron, nine-atom nido species. Instead, the surprise is found with the substantially D_{3h} symmetry of the isoelectronic Bi_9^{5+}, a result which may be related to the acidic, ionic environment necessary for the stability of this apolar cation, in contrast to the usual dipolar solvents which would stabilize the polar Sn_9^{4-}, etc. An appreciable elongation of the trigonal prism in Bi_9^{5+} relative to Ge_9^{2-} and $B_9H_9^{2-}$ may be a reflection of the accommodation of an additional pair of electrons in an approximately non-bonding orbital (21,22), a change which gives the cation a lower and more reasonable charge. According to CNDO calculations (10,12) the isolated Ge_9^{4-} and Sn_9^{4-} ions show very little ($\leq$ 0.5 eV) dependence of the atomization energy on configuration between the observed C_{4v} and the hypothetical D_{3h} limit, a result which led to a prediction of their fluctionality in solution.

Ralph Rudolph made important contributions to the question of fluctionality, and he provided virtually the only modern information on the solutions of the Zintl ions. Unfortunately the publications are at present limited to two communications (23,24), one conference proceedings (25), and a brief note on electrochemical generation of nine-atom species (26). Fluctionality of Sn_9^{4-} was established by the observation of a single NMR transition split by intramolecular coupling between ^{119}Sn and ^{117}Sn. A single ^{207}Pb resonance for Pb_9^{4-} and tin and lead resonances for all possible $Sn_{9-x}Pb_x^{4-}$ species were observed when a mixed alloy was extracted with en (without crypt). The $Sn_{9-x}Ge_x^{4-}$ family was also inferred from tin spectra. An important observation, consistent with our synthetic experiences, was that interanionic equilibrium in solution is very slow and these reactions occur principally at the surface of the solid alloy phases.

Rudolph and his coworkers also reported observation by ^{119}Sn and ^{117}Sn NMR of what was interpreted to be a novel, fluctional tetrahedral Sn_4^{2-} species when they used Na-Sn alloys poorer in tin, comparable to those used in the preparation of the highly insoluble $(cp\text{-}Na^+)_2Sn_5^{2-}$. A diamagnetic "tetrahedral" species

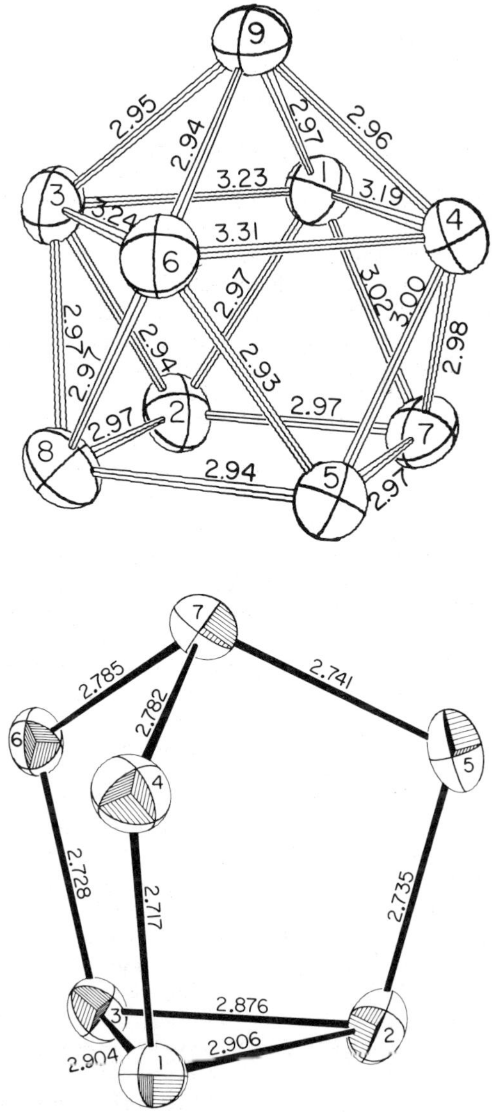

Figure 1. Structures of two homopolyatomic anions. Key: top, Sn_9^{4-} from crypt-Na^+ salt at 5° [$\sigma(d)$ = 0.004 to 0.005 Å] (12); and bottom, Sb_7^{3-}, from crypt-K^+ salt at −80° [$\sigma(d)$ = 0.002 Å] (14).

was interpreted theoretically in terms of a dynamic Jahn-Teller distortion of the 18-electron anion (27). A more-or-less tetrahedral but badly disordered Ge_4^{2-} which is presumably similar has been studied structurally (28), but the compound once thought to contain the analogous Sn_4^{2-} is now known to be a salt of $Sn_2Bi_2^{2-}$ (29).

An obvious absentee from the summary in Table I is Pb_9^{4-}, the first polymetal anion discovered. Many experiments involving Na-Pb and K-Pb reactions with crypt or its benzo derivative either in en or NH_3 have only given the Pb_5^{2-} salts or alloys as solid phases. The intermediates Pb_7^{4-} and Bi_5^{3-} reported by Zintl likewise have not been isolated, as well as the Sb_3^{3-} and Bi_3^{3-} ions which would be substantially more reduced than any species obtained so far from solution.

Electronic Requirements

The lack of homopolyatomic anions for elements to the left of group IV in Table I is noteworthy. Zintl reported no success with reactions of alkali metal alloys of the copper and zinc family elements and of thallium with liquid ammonia, and the generally stabilizing effect of crypt has not been evident in our own investigations of alloys of mercury and thallium. On the other hand, it is possible to isolate a white crypt-potassium gold compound from ammonia solutions at low temperatures which decomposes to elemental gold (+?) above about -10°C (30). Although the pseudohalide Au^- seems a likely candidate in the white solid, the decomposition seems unusual considering the high stabilities of KAu and the semiconducting CsAu (discussed in 31), unless the decomposition reaction actually involves reduction of the ligand.

The apparent instability of homopolyatomic anions of elements to the left of group IV (and of polyatomic cations to the left of group V) is thought to result from a deficiency of bonding electrons in species where the principal bonding is considered to originate largely from p-type orbitals. The electron deficient polyboranes, $B_nH_n^{2-}$, with which these clusters have some similarities in bonding (32), contain 2.17 to 2.4 p-electrons per boron atom for n between 12 and 5, respectively, and similar requirements apply to the homopolyatomic anions of germanium, tin and lead as well. The p-electron count per skeletal atom (e/M) shown in Figure 2 for most of the structurally known polyanions and cations of the fourth, fifth and sixth periods reveal that a relatively clear minimum of about 2.2 electrons per atom is necessary for stability in all, in spite of an expectation of weaker bonding for the heavier elements. All cluster and cage ions to the right of 3e/M in the Figure can be further described as containing ~2.2 - 3.0 skeletal bonding electrons per atom, with the remaining number relegated to the non-bonding lone pairs which are mainly important in determining the

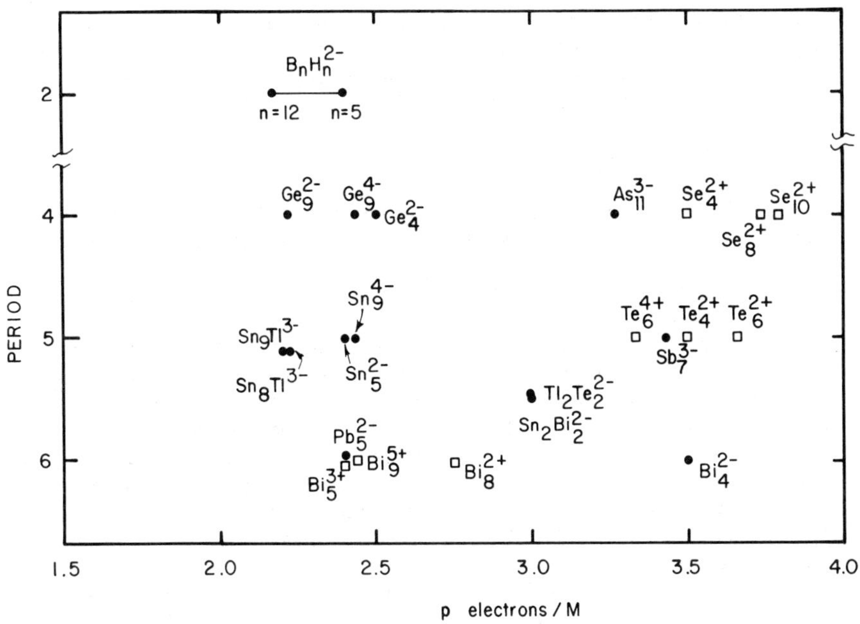

Figure 2. Number of skeletal p electrons per atom in cluster anions (□) and cations (●) as a function of period (33).

shape of the cluster (30). Many polychalcogenide anions with >4e/M are also in this category.

Heteropolyatomic Anions

A rather obvious means of incorporating the electron-poorer elements from groups earlier in the periodic table into clusters is then to combine them with electron-richer elements. Some results of synthetic and structural investigations of this sort involving ions with 2.2 - 3e per cluster element are listed in Table II together with data for three electron-richer mixed cluster anions.

Table II
Some Heteroatomic Examples of Polyatomic Anions

Anion	Symmetry	p electrons	Compare	Str. Ref.
$Sn_2Bi_2^{2-}$	~ T_d	12	P_4, Sb_4	29
$Pb_2Sb_2^{2-}$	~ T_d	12	P_4, Sb_4	14
$Tl_2Te_2^{2-}$	C_{2v}	12	-	30
$TlSn_8^{3-}$	C_{2v}(D_{3h}[a])	20	Ge_9^{2-}	33
$TlSn_9^{3-}$	C_{4v}(D_{4d}[a])	22	$B_{10}H_{10}^{2-}$	33
$As_2Se_6^{2-}$	~ C_{2h}	32	-	34
$HgTe_2^{2-}$	$D_{\infty h}$	10	HgX_2	35
$SnSe_4^{4-}$ [b]	T_d	22	SbS_4^{3-}	36

[a]Configuration neglecting heteroatom
[b]No crypt

The new anions nicely illustrate how the heteroatomic cluster anions enable one to sample electronic and structural configurations which are not (or can not) be achieved in the homoatomic regime. Thus although heavy metal tetrahedra are found in solid M^ISn and M^IPb phases there is no evidence that the anions Sn_4^{4-} or Pb_4^{4-} occur in solution, probably because of the high charge per atom. But the mixed species $Sn_2Bi_2^{2-}$ and $Pb_2Sb_2^{2-}$ shown in Figure 3 are quite stable and doubtlessly the alternate pairs $Sn_2Sb_2^{2-}$ and $Pb_2Bi_2^{2-}$ will be as well. Our studies have generally featured elements from different periods so that they can be distinguished by X-rays. In fact the two new "tetrahedral" clusters shown occur with the two types of atoms relatively uniformly disordered over the four inequivalent and well defined metal positions between crypt-K^+ cations. In

$Sn_2Bi_2^{2-}$ (top, Figure 3) the refined occupancies vary between 44 and 52% of a tin atom in the four positions but none is statistically different from 50%. The observed distances then represent the appropriate average of Sn-Sn, Sn-Bi and Bi-Bi bond lengths, the shorter observed distance corresponding to a somewhat greater tin population. With the antipode $Pb_2Sb_2^{2-}$ (bottom) the metal sites are more differentiated, the occupancies range from 33 to 73% antimony, and the anion exhibits an appropriately greater range of distances, the longer ones being associated with those atoms with the higher lead populations.

With increasing differences between the elements (III-VI rather than IV-V) the formation of a tetrahedron for a 20-electron species becomes unprofitable, as demonstrated by the isoelectronic $Tl_2Te_2^{2-}$. Its structure corresponds to a rhomboid folded by about 50° along the short Tl-Tl diagonal. The Tl-Tl distance (3.60 Å) is still too long to be considered indicative of a significant bond. Closure of this molecule to a tetrahedron would require a substantial charge flow from tellurium to thallium, unlikely for such a polar entity. The observed configuration is considered to arise from the square (D_{2h}) configuration through elongation to a rhomboid because of the larger nonbonding radius of tellurium together with some fold at thallium to maintain (mainly p-orbital) bonds with angles nearer 90°.

Figure 4 shows the first 10-atom 'naked' cluster isolated, $TlSn_9^{3-}$. This occurs as a substantially 50-50 composite with $TlSn_8^{3-}$ within a single interstice defined by the relatively large cations. In this case seven atoms from each $TlSn_x^{3-}$ cluster occur in single well-defined positions, only atoms 6 and 8 and the capping thallium atom shown being resolvable. (The heteroatom in $TlSn_8^{3-}$ overlaps Sn9 in $TlSn_9^{3-}$.) In fact the principal deviations from C_{4v} (or the ideal D_{4d} neglecting the thallium atom difference) for $TlSn_9^{3-}$ appears to arise because of coupling during the refinement of atom 6 and 8 as separate pairs, these ending up somewhat too close to one another in this ion and too far apart in $TlSn_8^{3-}$ (where atoms 2-3-6' and 1-4-8' define a tin trigonal prism). This finding further illustrates how restrictive the cavities among the crypt-M^+ cations may be for some configurations, although a fair number of disordered anion arrangements and approximately close-packed cations have also been encountered with other compositions. The thallium placement in both $TlSn_8^{3-}$ and $TlSn_9^{3-}$ is at variance with that developed for the heteroboranes. In both cases the electron-poorer element (Tl) is found to substitute at the lower order and electron-<u>richer</u> vertex, the charge distribution being inferred from CNDO calculations for the equivalent homoatomic clusters (33).

The $TlSn_9^{3-}$ cluster can be viewed as originating through capping of the open face of Sn_9^{4-} (Figure 1) by Tl^+ while its companion $TlSn_8^{3-}$ is isoelectronic with the known Ge_9^{2-} (Table I). No evidence has yet been found for the species Sn_{10}^{2-} or

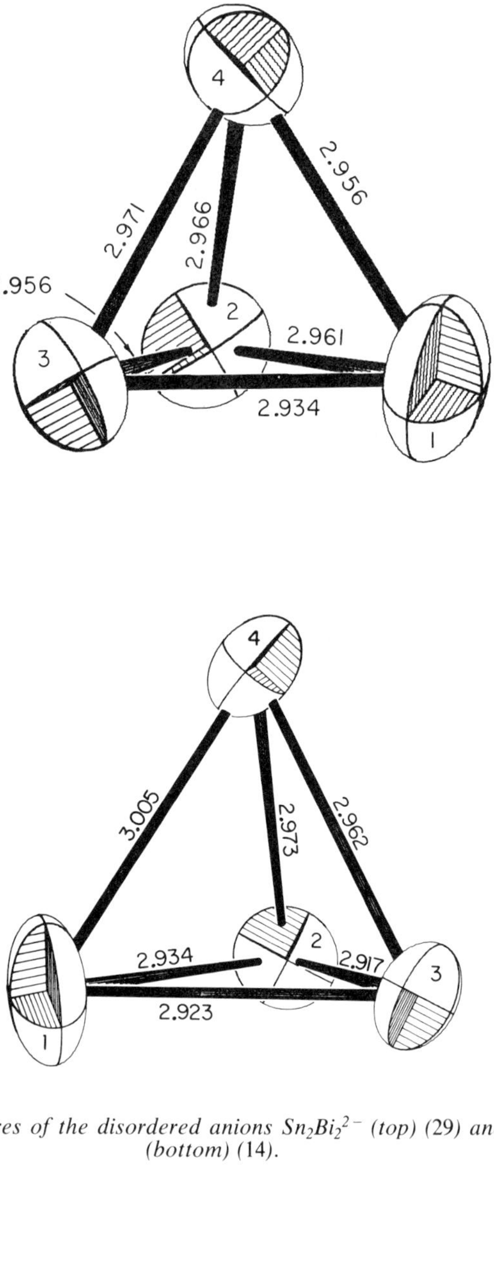

Figure 3. Structures of the disordered anions $Sn_2Bi_2^{2-}$ (top) (29) and $Pb_2Sb_2^{2-}$ (bottom) (14).

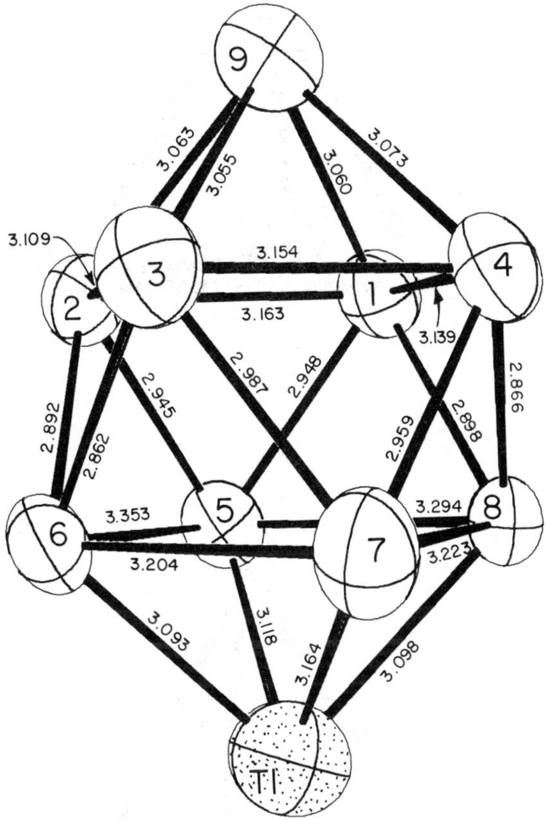

Figure 4. The $TlSn_9^{3-}$ anion in $(cp\text{-}K^+)_6(TlSn_9^{3-}\text{-}TlSn_8^{3-}) \cdot 2en$ (33).

Pb_{10}^{2-} which would be isoelectronic with $TlSn_9^{3-}$, perhaps because only two cations would provide an inadequate separation of these large anions.

The remainder of the mixed anion examples in Table II are only a forecast of the variety possible for electron-richer anions. Rudolph, et al. (24), utilized ^{119}Sn and ^{125}Te NMR on solutions of Na-Sn-Te alloys in en to infer the presence of the tetrahedral, tin-centered $SnTe_4^{2-}$ which is, therefore, not a polyhedral cluster anion. The $SnSe_4^{4-}$ analog cited is the only one which has been structurally verified. Other mixed anions involving elements from different groups have been observed by Rudolph and coworkers by tin NMR, viz., Sn_9Sb^{x-}, where x is thought to be either 1 or 3, and a fluctional $TlSn_8^{5-}$, the charge assignment in the latter being based on the analysis of the sodium bis-en salt (24). The latter is thus different from the $TlSn_8^{3-}$ known structurally (Table II).

Prior to any work on heteroatom clusters the notion was expressed (20) that heteroatom placement within the polyatomic clusters would lead to a decrease in delocalization and bonding and thence stability. Although this may lessen stability the substitution clearly does not preclude it. Furthermore, many of the likely polyhedra already have inequivalent atom positions, the 5, 7, 9 and 10 atom examples already considered here for example, and mixed species especially with elements from different groups may be quite stable within the discrimination provided by inequivalent positions. Even the nominally equivalent atom positions in a tetrahedron can obviously accommodate substantial differences. Additional examples of mixed element polycations are certainly to be expected. An inadequate foresight was revealed in a review of polycations (20) written for a 1974 award symposium, about one year before the crypt discoveries, by the expectation that polycations should be more stable than polyanions for the metallic elements. In hindsight, metallic behavior is a property of the dense solid state and has little to do with the stability of small clusters where electronic and geometric factors are far more important.

The synthesis of solid derivatives homo- and heteroatomic polyanions has to date principally featured the simplest of techniques to obtain generally the least soluble and most stable of the ions possible from either en or NH_3. There is no question that many other species exist in solution if not as solids, as evidenced by Zintl's results, numerous color changes observed during reactions of alloys with crypt solutions, several semi-quantitative analyses of either unsatisfactory crystals or solutions formed in mixed metal reactions (14,37), and new NMR signals from unknown species in a wide variety of systems (23,24,38). The solution chemistries alone deserve a great deal more attention.

A Not-So-Naked Cluster

Finally, a particular surprise from this work has been $(cp\text{-}K^+)_3(KSn_9^{3-})$, the structure of which appears to represent a partial transition to an intermetallic-like bonding. The compound which was isolated in respectable yield from reactions of K-Hg-Sn alloys with crypt in en exhibits infinite chains K(4)-Sn_9-K(5)-Sn_9-K(4) running diagonally through the triclinic cell, K(4) and K(5) being located on centers of symmetry at 0,0,0 and $^1/_2$,$^1/_2$, $^1/_2$. Six crypt-K^+ cations fill the remainder of the cell (R = 0.075, R_w = 0.090 for data to 2Θ = 45°) (37).

The independent anion unit shown in Figure 5 contains two surprises. First, the potassium atoms have four tin neighbors at 3.55 to 3.74 Å through bridging opposite edges on the waist of the Sn_9 unit, with two more tin atoms at greater distances of 4.11 or 4.22 Å to produce a distorted trigonal antiprism of tin about each potassium. There is no evidence (± 0.5e/$Å^{-3}$) for en groups coordinated to either of the potassium atoms. Second, the distances in the Sn_9^{4-} group are scarcely altered by the presence of the two potassium atoms, as can be as seen by comparison of these results with those for the 'naked' Sn_9^{4-} shown in Figure 1.

Judging from K_4Sn_4 where potassium has 6 to 8 tin neighbors in the range 3.70 to 3.85 Å (maximum deviation = 0.03 to 0.05 Å) (39), the potassium-tin interactions found here appear normal except for the short pair at 3.55 Å about K(4). Within the Sn_9 portion of the structure, the similar Sn-Sn distances in all but the square capped face average 2.964 (5) Å in this compound compared with 2.966 (6) Å in Sn_9^{4-}. The structural data were collected at -100 and 5°, respectively, so that some expansion of the cluster in KSn_9^{3-} would probably be seen in an isothermal comparison. In addition, there appears to be a perceptible distortion of the cluster in the new potassium salt in that Sn-Sn distances where both atoms have potassium neighbors are longer by an average of 0.027 Å. Although the charge transfer to potassium is seemingly not large enough to alter the tin cluster very much, perhaps it is enough to eliminate the (relatively weak) tendency for potassium to coordinate en. The packing of the crypt-K^+ cations can not be a significant factor in preventing all coordination of solvent since the shortest distance between these potassium atoms and the nearest light atom in crypt is 4.27 Å. With these unusual circumstances it is somewhat reassuring to find another example of association of a cluster with an alkali metal cation. In Rb_3As_7·3en the rubidium atoms all bridge edges of the anion, with the solvent coordinated to the alkali metal only on the outside of the Rb_3As_7 'complex' (40).

Acknowledgment

This research was supported by the Office of Basic Energy Sciences, Materials Sciences Division. The Ames Laboratory is operated for the U. S. Department of Energy by Iowa State University under contract No. W-7405-Eng-82.

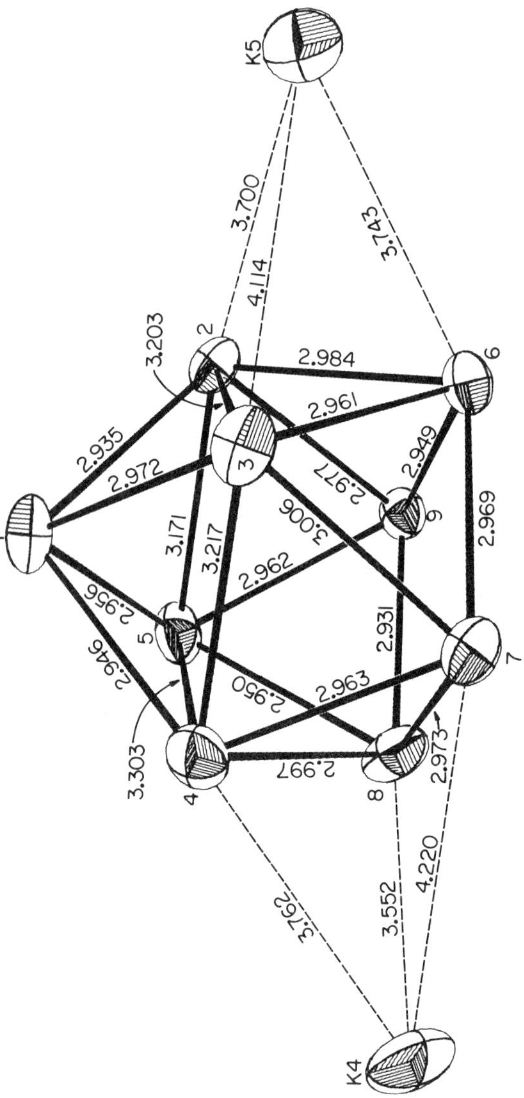

Figure 5. The anion chain in $(cp\text{-}K^+)_3(KSn_9^{3-})$ [$\sigma(d)$ = 0.002 *to* 0.003 Å] (37).

Literature Cited

1. Joannis, A. C. R. Hebd. Seances Acad. Sci. 1891, 113, 745; 1892, 114, 587.
2. Smyth, F. H. J. Am. Chem. Soc. 1917, 39, 1299.
3. Kraus, C. A. J. Am. Electrochem. Soc. 1924, 45, 175.
4. Zintl, E.; Goubeau, J; Dullenkopf, W. Z. Phys. Chem., Abt. A 1931, 154, 1.
5. Zintl, E.; Harder, A. Z. Phys. Chem., Abt. A 1931, 154, 47.
6. Zintl, E.; Dullenkopf, W. Z. Phys. Chem., Abt. B 1932, 16, 183.
7. Kummer, D.; Diehl, L. Angew. Chem., Int. Ed. Engl. 1970, 9, 895.
8. Diehl, L.; Khodadedeh, K.; Kummer, D.; Strahle, J. Chem. Ber. 1976, 109, 3404.
9. Corbett, J. D.; Adolphson, D. G.; Merryman, D. J.; Edwards, P. A.; Armatis, F. J. J. Am. Chem. Soc. 1975, 97, 6267.
10. Belin, C. H. E.; Corbett, J. D.; Cisar, A. J. Am. Chem. Soc. 1977, 99, 7163.
11. Corbett, J. D.; Belin, C. H. E., unpublished research.
12. Corbett, J. D.; Edwards, P. J. Am. Chem. Soc. 1977, 99, 3313.
13. Edwards, P.; Corbett, J. D. Inorg. Chem. 1977, 16, 903.
14. Critchlow, S. C.; Corbett, J. D., to be published.
15. Cisar, A.; Corbett, J. D. Inorg. Chem. 1977, 16, 2482.
16. Belin, C. H. E. J. Am. Chem. Soc. 1980, 102, 6036.
17. Adolphson, D. G.; Corbett, J. D.; Merryman, D. J. J. Am. Chem. Soc. 1976, 98, 7234.
18. Teller, R. G.; Krause, L. J.; Haushalter, R. C. Inorg. Chem. 1983, 22, in press.
19. von Schnering, H.-G. Angew Chem., Int. Ed. Engl. 1981, 20, 33.
20. Corbett, J. D. Prog. Inorg. Chem. 1976, 21, 129.
21. King, R. B. 1981, private communication.
22. Burns, R. C.; Gillespie, R. J.; Barnes, J. A.; McGlinchey, M. J. Inorg. Chem. 1982, 21, 806.
23. Rudolph, R. W.; Wilson, W. L.; Parker, F.; Taylor, R. C.; Young, D. C. J. Am. Chem. Soc. 1978, 100, 4629.
24. Rudolph, R. W.; Wilson, W. L.; Taylor, R. C. J. Am. Chem. Soc. 1981, 103, 2480.
25. Rudolph, R. W.; Taylor, R. C.; Young, D. C. in "Fundamental Research in Homogeneous Catalysis", M. Tsutsui, Ed.; Plenum Press: New York, 1979, p. 997.
26. Pons, B. S.; Santure, D. J.; Taylor, R. C.; Rudolph, R. W. Electrochem. Acta 1981, 26, 365.
27. Rothman, M. J.; Bartell, L. S.; Lohr, L. L. J. Am. Chem. Soc. 1981, 103, 2482.
28. Critchlow, S. C.; Corbett, J. D. J. Chem. Soc., Chem. Commun. 1981, 236.
29. Critchlow, S. C.; Corbett, J. D. Inorg. Chem. 1982, 21, 3286.

30. Burns, R. C.; Corbett, J. D. J. Am. Chem. Soc. 1981, 103, 2627.
31. Hensel, F. Adv. Phys. 1979, 28, 589.
32. Corbett, J. D. Inorg. Chem. 1968, 7, 198.
33. Burns, R. C.; Corbett, J. D. J. Am. Chem. Soc. 1982, 104, 2804.
34. Belin, C. H. E.; Charbonnel, M. M. Inorg. Chem. 1982, 21, 2504.
35. Burns, R. C.; Corbett, J. D. Inorg. Chem. 1981, 20, 4433.
36. Krebs, B.; Hinter, H.-V. Z. Anorg. Allg. Chem. 1980, 462, 143.
37. Burns, R. C.; Corbett, J. D., to be published.
38. Rudolph, R. W. 1980, private communication.
39. Hewaidy, I. F.; Busmann, E.; Klemm, W. Z. Anorg. Allg. Chem. 1964, 328, 283.
40. Honle, W.; von Schnering, H.-G. 1983, to be published.

RECEIVED April 8, 1983

7

Multihapto Bonding Between Main Group Elements and Carbocyclic Ligands

An Approach to the Bonding in Main Group Cluster Compounds

S. G. BAXTER, A. H. COWLEY, and J. G. LASCH

The University of Texas at Austin, Department of Chemistry, Austin, TX 78712

MNDO calculations have been performed on a number of main-group capping moieties for C_3H_3, C_4H_4, C_5H_5, and C_6H_6 carbocyclic rings. Perhapto bonding is predicted when the number of interstitial electrons (comprising ring-π plus relevant main-group fragment electrons) totals six. Some cases of intermediate hapticity are discussed in the context of the C_5H_5 ring.

The discovery of the transition metal sandwich molecules ferrocene and dibenzenechromium ranks as one of the more important chemical discoveries over the past three decades. The chemistry of these and other d- and f- block carbocyclic π-complexes has been investigated intensively in the ensuing years. Significant progress has also been made toward understanding the electronic structures and patterns of stability of these interesting compounds (1). However, compared with the d- and f-block elements, much less is known about the multihapto interaction of the main-group elements with carbocyclic ligands. Minkin and Minyaev (2) have suggested that annulene-cap perhapto bonding will be favored when the total number of ring-π plus main-group moiety electrons is equal to eight, while Schleyer *et al.*(3) favor an optimum number of six interstitial electrons (4).

Several main-group π-complexes have been investigated using various levels of theory. Thus, C_5H_5Li (5) and C_5H_5BeH (6) have been computed using *ab initio* methods, and Dewar and Rzepa (7) have explored the interaction of BeX moieties with cyclopentadienyl, indenyl, and fluorenyl groups. The MNDO method has also been employed for the investigation of various carbocyclic beryllium derivatives, (8) and the PRDDO method has been utilized in conjunction with C_5H_5BeR compounds (9). The tin cation $[C_5H_5Sn]^+$ has been studied with the EHMO method, (10) and Hoffmann

0097–6156/83/0232–0111 $06.00/0

et al. (11) have examined the motion of a $[CH_2]^{2+}$ fragment across a $[C_5H_5]^-$ ring.

The present contribution represents an attempt to develop further a model for the multihapto bonding of main-group element fragments to carbocyclic rings, $(CH)_n$ (n=3-6) using the MNDO program system (12). One of the advantages of this method is that it permits study of the capping of carbocyclic rings by the heavier main-group elements. Emphasis is placed basically on two themes: (i) whether a six- or eight-electron counting procedure is more appropriate, and (ii) whether conformity with other cluster electron counting procedures is achieved. It is on the latter point that Ralph Rudolph made one of his seminal contributions to inorganic chemistry. Independently he (13) and Wade (14) arrived at a method of counting skeletal electrons which provided major new insights into the systematics of main-group and transition metal cluster chemistry.

Computational Procedures

All computations were carried out using the MNDO method (12) and published parameters (15). For the cyclic structures, transits of a perpendicular main-group moiety across the various rings generate η^n structures (n=1-6) as shown below.

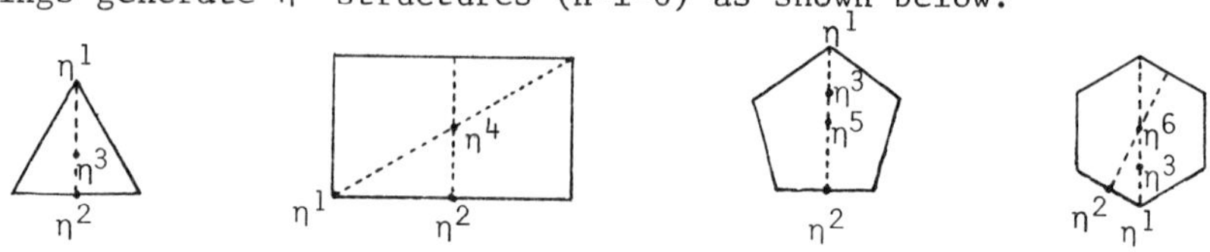

This approach has been discussed previously for the cyclopentadienyl ring and has been referred to as a haptotropic search (11,6). The geometry definitions were as follows:

Perhapto Structures. In all structures, the carbon atoms of the $(CH)_n$ ring (n=3-6) were constrained as a regular polygon, but in each calculation the C-C bond distances were relaxed. To ensure a perhapto final structure, the position of the central atom, M, was minimized along the principal axis of the ring. The hydrogens were held symmetry - equivalent (*i.e.* all C-H bond lengths and all C-C-H bond angles equal), but were allowed to bend out of the plane of the ring.

Dihapto Structures. For these calculations, the ring carbon atoms were frozen into the geometries obtained in the perhapto calculations. Again, symmetry-equivalent hydrogen atoms were constrained to have equal C-H bond lenths and C-C-H bond angles. However, no torsion angle limitations were imposed. The central atom was maintained in a dihapto configuration by relaxation of its position on a line perpendicular to the ring plane and which bisected a C-C bond.

Monohapto Structures. As in the case of the dihapto structures, the geometries of the ring carbons were constrained to those emerging from the perhapto calculations, and symmetry-equivalent hydrogens were forced to maintain equal C-H bond lengths and C-C-H bond angles. Monohapticity was imposed by minimizing the position of the central atom along a line perpendicular to the ring plane and passing through a ring carbon.

Structures with MH Moieties. For perhapto structures the hydrogen was confined to the same line as the central atom. The dihapto and monohapto calculations also used input structures with the hydrogen on the same line used to constrain the central atom, but here the position of the hydrogen atom was totally relaxed.

Structures with MH_2 Moieties. These were handled on a geometry-optimized basis.

Qualitative Considerations

Due to the availability of appropriate organometallic reagents, by far the largest number of annulene complexes of the main-group elements involve the cyclopentadienyl group. Let us start the discussion, therefore, by considering qualitatively the π-type coordination of a C_5H_5 ring to a main-group element, M. As shown in Figure 1a, bonding interactions in C_{5v} symmetry are expected between the valence s orbital of M and the totally symmetric π-MO of C_5H_5, and between the degenerate np_x and np_y AO's of M and the e_1 MO of C_5H_5. The $2a_1$ MO results from mixing of the antibonding component of the $1a_1$ MO and a bonding MO which results from interaction between the valence p_z orbital and the a_1 ring MO. The relative energies of the e_1 and $2a_1$ MO's are going to depend on e.g. the energies of the valence s and p orbitals of M.

Somewhat similar diagrams can be constructed for the interaction between a cyclopentadienyl group and ligated main-group moieties, ML_n. That for a $C_5H_5\cdots\cdot MH$ interaction is illustrated in Figure 1b.

Obviously, eight electrons can be accommodated in either the $(\eta^5-C_5H_5)M$ or $(\eta^5-C_5H_5)MH$ bonding scheme. However, the question of whether a six- or eight-electron interstitial electron count is more valid will depend on (a) whether the $2a_1$ MO is occupied, and (b) if the $2a_1$ MO is occupied whether it involves significant ring-M interaction.

Realizing that the basic pattern of a and e ligand π MO's persists throughout the carbocyclic ring systems, C_nH_n (n=3-6) (16), it is clear that the interaction of an s,p basis set of a main-group element with C_3H_3, C_4H_4, and C_6H_6 rings should produce $1a_1$, $1e_1$, and $2a_1$ MO's in a very similar fashion to the cyclopentadienyl systems illustrated in Figure 1.

Five-Membered Ring Systems. In terms of haptotropic searches,

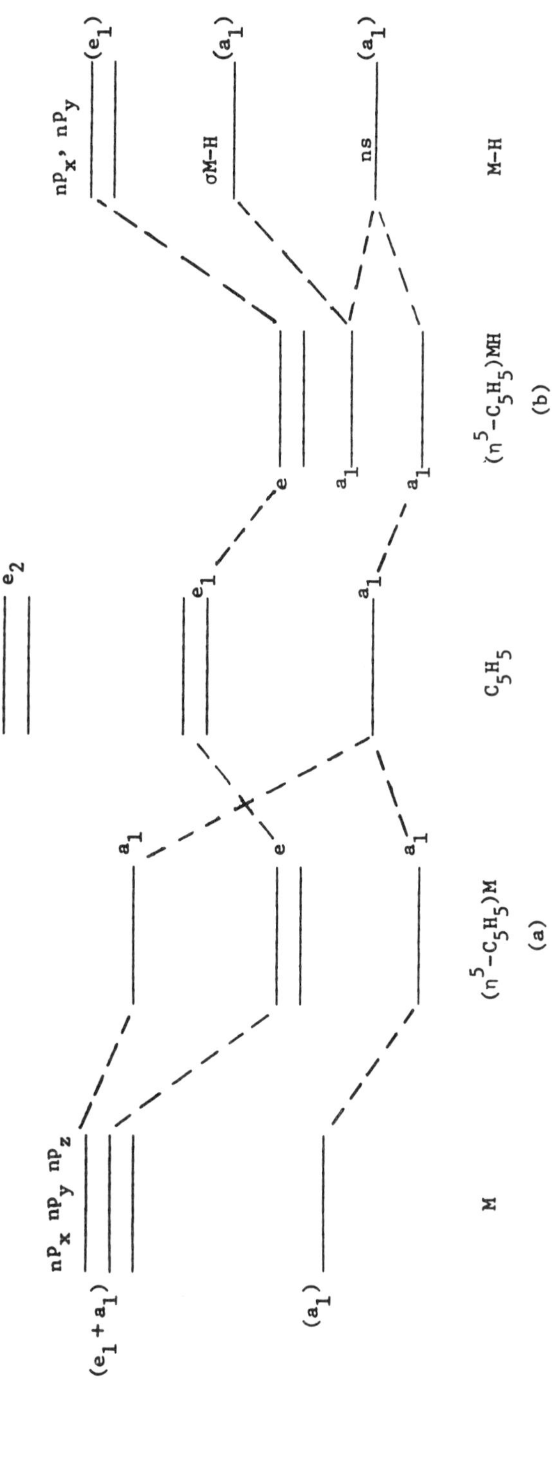

Figure 1. Qualitative scheme for the pentahapto interaction of C_5H_5 π-orbitals with a main group element M (left) or an M–H σ-bond (right).

both C_5H_5B and C_5H_5Al adopt η^5 ground state geometries (Table I). Furthermore, the η^5 preference prevails for both molecules when a complete geometry optimization is carried out. Only minimal distortion of the rings takes place during the geometry optimization. Both C_5H_5B and C_5H_5Al are unknown molecules; however, the heavier congeners C_5H_5In and C_5H_5Tl have been known for several years. Although C_5H_5In and C_5H_5Tl are polymeric in the solid state (17), these molecules adopt η^5 structures in the vapor state (18). Although C_5H_5B and C_5H_5Al both adopt η^5 geometries, it is interesting to note that the sequence of MO's differs in the two molecules; in C_5H_5B the HOMO is of e symmetry, while that of the aluminum analog is a_1. A NOCOR MO calculation (19) on C_5H_5Tl indicates that the HOMO is of a_1 symmetry; however, UV PES spectral data (20) and Xα scattered-wave calculations (21) for C_5H_5In and C_5H_5Tl suggest that the HOMO is of e symmetry.

We now turn to some charged systems. The anion $[C_5H_5Be]^-$ is isoelectronic with C_5H_5B and, like the boron compound, exhibits a preference for the η^5 geometry. This preference persists when the geometry is completely optimized. The calculations on $[C_5H_5Si]^+$ were undertaken because the heavier congeneric cation $[Me_5C_5Sn]^+$ is known and has been found to exhibit a pentahapto structure on the basis of X-ray crystallography (10). The MNDO haptotropic search indicates a minimum at η^5; moreover, the pentahapto structure persists with only minor changes of energy when the geometry is optimized. The $[C_5H_5Si]^+$ cation has also been investigated by _ab initio_ methods (3c).

The calculations on $[C_5H_5BH]^+$ were undertaken because boron cations of the type $[Me_5C_5BX]^+$ (X=Cl, Br, I) are known (22), and spectroscopic evidence indicates that the BX^+ moiety is pentahapto-bonded to the Me_5C_5 ring. Moreover, C_5H_5BeH, which is isoelectronic with $[C_5H_5BH]^+$ has been shown to possess an η^5 geometry (23). Our MNDO calculations on $[C_5H_5BH]^+$ imply that the η^2 and η^5 geometries are very close in energy. Previously, it has been suggested (24) that the addition of boron Lewis Acids to C_5H_5In causes a change from η^5 to η^1 attachment.

Six or Eight Interstitial Electrons?

For the systems considered in Table I, all eight electrons would be counted according to the Minkin and Minyaev model (2). In the model of Schleyer _et al._, (3) a main-group element lone pair of C_5H_5B, C_5H_5Al, $[C_5H_5Be]^-$, and $[C_5H_5Si]^+$ and the B-H σ-bond of $[C_5H_5BH]^+$ would be excluded from the count on the basis that they are not interstitial electrons. Xα-SW calculations on e.g. $(\eta^5\text{-}C_5H_5)In$ and $(\eta^5\text{-}C_5H_5)BeX$ support the views of Schleyer _et al._ (3a). Thus, the $5a_1$ MO of $(\eta^5\text{-}C_5H_5)In$, which is predominantly lone pair character, and the $4a_1$ MO of $(\eta^5\text{-}C_5H_5)BeX$, which is the Be-X σ-bond, feature only minor contributions from the C_5H_5 ring.

It should also be noted that a six-electron rule affords conformity with the electron counting procedures of Rudolph (13)

Table I. Heats of Formation (kcal/mol), Relative Energies (kcal/mol), and η^5 Geometries (Å) for C_5H_5 Compounds.

COMPOUND	HAPTICITY	ΔH_f	(RELATIVE ENERGY	η^5 GEOMETRY
C_5H_5B	η^1	112.21	(8.29)	B-C = 1.904
	η^2	112.83	(8.91)	C-C = 1.446
	η^5	103.92	(0.00)	C-H = 1.081
	optimized	103.90		
C_5H_5Al	η^1	76.59	(5.61)	Al-C = 2.215
	η^2	76.11	(5.13)	C-C = 1.442
	η^5	70.98	(0.00)	C-H = 1.082
	optimized	70.97		
$[C_5H_5Be]^-$	η^1	75.53	(6.69)	Be-C = 2.188
	η^2	75.01	(6.17)	C-C = 1.439
	η^5	68.84	(0.00)	C-H = 1.083
	optimized	68.84		
$[C_5H_5Si]^+$	η^1	213.02	(23.13)	Si-C = 2.089
	η^2	205.05	(15.16)	C-C = 1.453
	η^5	189.89	(0.00)	C-H = 1.086
	optimized	189.89		
$[C_5H_5BH]^+$	η^1	277.97	(9.79)	B-C = 1.784
	η^2	268.18	(0.00)	C-C = 1.471
	η^5	269.24	(1.06)	C-H = 1.086
	optimized	269.23		B-H = 1.161

and Wade (14). According to these rules, all the compounds in Table I are predicted to adopt nido structures since each possesses n+2 skeletal electron pairs. Each CH group contributes three skeletal electrons, hence only one valence electron is required from B, Al, Be^-, Si^+, or BH^+ to achieve a total of six electrons.

Consequences of Increasing the Number of Valence Electrons on the Main-Group Fragment

All the systems considered in Table I feature a main-group fragment electron count of three, only one electron of which is involved in interstitial bonding. We now explore the consequences of increasing the total electron count of the main-group fragment to five electrons. Two examples of this electron count are represented by $C_5H_5MH_2$ (M=B, Al). For both molecules the minimum energy geometry is η^1 and the M····C(2) and M····C(5) distances

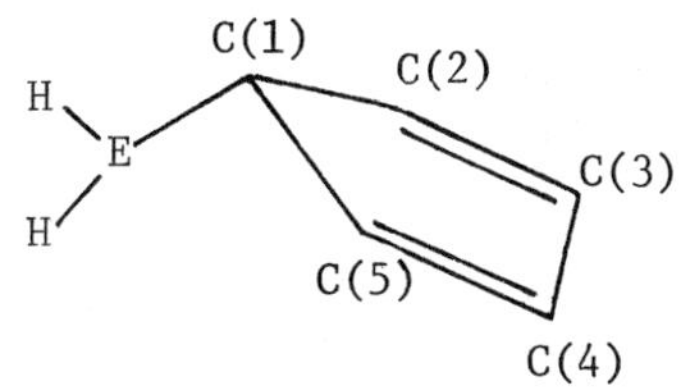

E = B		E = Al	
B-C(1)	1.55 Å	Al-C(1)	1.84 Å
C(1)-C(2)	1.52 Å	C(1)-C(2)	1.51 Å
C(2)-C(3)	1.36 Å	C(2)-C(3)	1.37 Å
C(3)-C(4)	1.47 Å	C(3)-C(4)	1.47 Å
C(4)-C(5)	1.36 Å	C(4)-C(5)	1.37 Å
C(1)-C(5)	1.53 Å	C(1)-C(5)	1.52 Å

exceed the sum of covalent radii for B and C (1.57 Å) or Al and C (~2.00 Å). Previous ab initio work (25) had indicated a minimum at η^2; however the barriers between η^2, η^3, and η^5 structures were found to be small. An EHMO investigation (11) of the system $[C_5H_5]^-\cdots[CH_2]^{2+}$ also indicated an η^2 minimum but no minimum was found at the η^3 geometry. The experimental data for a number of Group IIIA compounds of the type (C_5R_5)MXY (R=H, Me; X, Y=alkyl, C_5H_5, Cl) have been assembled in Table II. Collectively, these data indicate that the energies of η^1, η^2, and η^3 geometries are rather close. Similar conclusions have emerged from MNDO calculations (35) on phosphenium ions of the type $[(Me_5C_5)(R)P]^+$, namely (i) that the global minimum is η^2, (ii) η^5 and η^3 structures do not correspond to minima, and (iii) the barrier to circumannular migration of the RP moiety in the η^2 structures (via an η^1 intermediate) is very small (<2 kcal/mol).

A count of five main-group valence electrons can also be obtained by e.g. adding H^- to the silicon cation, $[C_5H_5Si]^+$. This system exhibits a local minimum with η^2 attachment of the SiH moiety to the C_5H_5 ring; however, the global minimum corresponds to the known molecule silabenzene (36).

Table II. Some Cyclopentadienyl and Pentamethylcyclopentadienyl Main Group Compounds with Intermediate Hapticities.

SYSTEM	HAPTICITY	METHOD	REFERENCE
$C_5H_5BEt_2$	η^1	NMR	26
$C_5H_5AlMe_2$	η^2 or η^3	ED (gas)	27
$C_5H_5MR_2$ (R = Me, Et; M = Al, Ga, In)	η^2 or η^3	vibrational spectroscopy	28
$[C_5H_5GaMe_2]_n$	η^1	X-ray	29
$[C_5H_5AlMe_2]_n$	η^1	X-ray	30
$[(Me_5C_5)Al(Me)Cl]_2$	η^3	X-ray	31
$[(Me_5C_5)Al(iBu)Cl]_2$	between η^2 and η^3	X-ray	32
$[(C_5H_5)_3In]_n$	η^1	X-ray	33
$[Me_5C_5Sn]^+\cdot$Pyridine	η^2		34

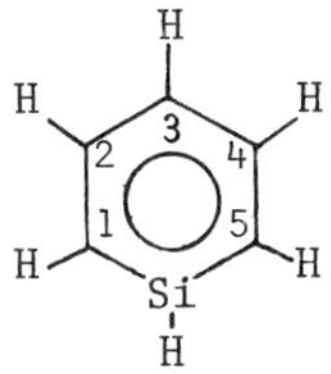

Si-C(1)	1.72 Å
Si-C(5)	1.72 Å
C(1)-C(2)	1.40 Å
C(2)-C(3)	1.41 Å
C(3)-C(4)	1.41 Å
C(4)-C(5)	1.40 Å

A similar observation was made with the isoelectronic system C_5H_5P, namely an η^3 local minimum but with the known molecule, phosphabenzene (37) as the global minimum.

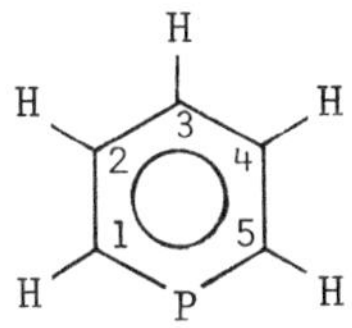

P-C(1)	1.64 Å
P-C(5)	1.64 Å
C(1)-C(2)	1.40 Å
C(2)-C(3)	1.41 Å
C(3)-C(4)	1.41 Å
C(4)-C(5)	1.40 Å

Further increases in the number of electrons associated with the main-group fragment are realized in the halides, C_5H_5X (X=F, Cl). The global minimum for both molecules is η^1 and this geometry is strongly preferred over any multi-hapto structure. This conclusion is in agreement with previous theoretical work which has shown that the preferred geometry for CH_3, SiH_3, or Me_3Si on a C_5H_5 ring is monohapto (11,38) and also with an X-ray crystal structure of $(\eta^1\text{-}Me_5C_5)SiCl_3$ (39) and NMR data on the series of molecules, $C_5H_5MMe_3$ (M=Si, Ge, Sn) (40).

Three-, Four-, and Six-Membered Ring Systems

The following perhapto structures emerge from haptotropic searches (see also Table III).

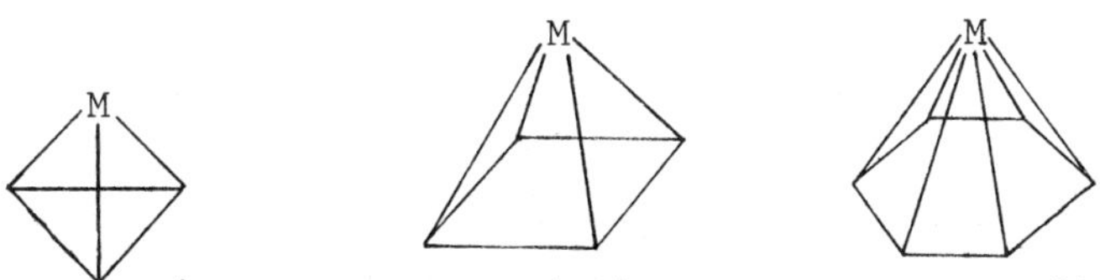

η^3: M = N, P, S^+, SiH η^4: M = N^+, P^+, Si, B^- η^6: M = Si^{2+}

Note that each perhapto structure is consistent with the six-electron rule of Schleyer et al. (3) in the sense that one pair of electrons on each main-group fragment is not involved in interstitial bonding.

None of the C_3H_3 systems, $(\eta^3\text{-}C_3H_3)N$, $(\eta^3\text{-}C_3H_3)P$, $[(\eta^3\text{-}C_3H_3)S]^+$, or $(\eta^3\text{-}C_3H_3)SiH$ is, to our knowledge, known experimentally. However, realizing that the fragments N, P, S^+, and SiH

Table III. Heats of Formation (kcal/mol), Relative Energies (kcal/mol), and Perhapto Geometries (Å) for C_nH_n Compounds (n = 3, 4, and 6).

COMPOUND	HAPTICITY	ΔH_f	(RELATIVE ENERGY)	PERHAPTO GEOMETRY
$(C_3H_3)N$	η^1	211.11	(69.44)	N-C = 1.502
	η^2	162.26	(20.59)	C-C = 1.503
	η^3	141.67	(0.00)	C-H = 1.065
$(C_3H_3)P$	η^1	147.29	(54.16)	P-C = 1.773
	η^2	106.49	(13.36)	C-C = 1.505
	η^3	93.13	(0.00)	C-H = 1.065
$(C_3H_3)S^+$	η^1	337.02	(0.00)	S-C = 1.837
	η^2	314.50	(3.96)	C-C = 1.493
	η^3	310.54	(26.48)	C-H = 1.075
$(C_3H_3)SiH$	η^1	168.72	(57.41)	Si-C = 1.789
	η^2	128.66	(17.35)	C-C = 1.547
	η^3	111.32	(0.00)	C-H = 1.070
				Si-H = 1.432
$[(C_4H_4)N]^+$	η^1	393.45	(45.92)	N-C = 1.617
	η^2	361.37	(13.84)	C-C = 1.498
	η^4	347.53	(0.00)	C-H = 1.084
$[(C_4H_4)P]^+$	η^1	315.58	(34.09)	P-C = 1.911
	η^2	290.33	(8.84)	C-C = 1.492
	η^4	281.49	(0.00)	C-H = 1.080
$(C_4H_4)Si$	η^1	87.00	(29.46)	Si-C = 1.966
	η^2	63.92	(6.38)	C-C = 1.484
	η^4	57.54	(0.00)	C-H = 1.073
$[(C_4H_4)B]^-$	η^1	122.79	(15.59)	B-C = 1.766
	η^2	110.73	(3.53)	C-C = 1.478
	η^4	107.20	(0.00)	C-H = 1.071
$[(C_6H_6)Si]^{2+}$	η^1	488.78	(10.61)	Si-C = 2.219
	η^2	487.72	(9.55)	C-C = 1.435
	η^6	478.17	(0.00)	C-H = 1.101

are isolobal (41) with CH, we note that tetrakis (t-butyl)tetrahedrane (which can be regarded as an (η^3-C_3H_3)CH derivative) has been prepared (42). Moreover, (η^3-C_3H_3)P can be regarded as one of three possible (but as yet unsynthesized) clusters of general formula $(CH)_xP_{4-x}$ (x=1, 2, 3) which are intermediate between P_4 and tetrahedrane.

None of the capped cyclobutadiene molecules or ions is known experimentally. Moreover, in an ab initio MO study of Be, BH, and CH^+ as capping moieties for C_4H_4, Jemmis and Schleyer (3a) have calculated that the corresponding five-membered planar structures are more stable. A similar situation may obtain in the present work. Nevertheless, it is worth remarking that the carbocation $[(CH)_5]^+$ has been investigated both theoretically (43) and experimentally (44) and found to adopt a structure in which a CH^+ fragment is η^4-bonded to a C_4H_4 ring.

Jemmis and Schleyer (3a) have found previously that Li^+, LiH, and BeH^+ are capable of binding in the hexahapto fashion to a benzene ring. On the other hand, no minimum could be discerned for hexahapto binding of Be to C_6H_6 utilizing the STO-3G basis set. Our MNDO haptotropic search indicates that Si^{2+} prefers to bond to C_6H_6 in the hexahapto manner. This conclusion is relevant to highly interesting experimental work by Amma and co-workers (45) which had demonstrated that Sn^{2+} is hexahapto bonded to C_6H_6 in the complexes (η^6-C_6H_6)$Sn(AlCl_4)_2 \cdot C_6H_6$ and (η^6-C_6H_6)$SnCl(AlCl_4)$.

Acknowledgment

The authors are grateful to the Petroleum Research Fund for generous financial support.

Literature Cited

1. For reviews, see (a) Mingos, D.M.P. Adv. Organomet. Chem. 1977, 15, 1; (b) Haaland, A. Acc. Chem. Res. 1979, 12, 415.
2. Minkin, V. I.; Minyaev, R. M. Zh. Org. Khim. 1979, 15, 225, 1569.
3. (a) Jemmis, E. D.; Schleyer, P. v. R. J. Am. Chem. Soc. 1982, 104, 4781. For related papers, see (b) Collins, J. B.; Schleyer, P. v. R. Inorg. Chem. 1977, 16, 152 (c) Krogh-Jespersen, K.; Chandrasekhar, J.; Schleyer, P. v. R. J. Org. Chem. 1980, 45, 1608.
4. Interstitial electrons have been defined (3b) as electrons which bind π-bonding ligands to a central atom.
5. (a) Janoschek, R.; Diercksen, G.; Preuss, H. Int. Quantum Chem. Symp. 1967, 1, 205 (b) Alexandratos, S.; Streitwieser, A., Jr.; Schaefer, H. F., III J. Am. Chem. Soc. 1976, 98, 7959.
6. Jemmis, E. D.; Alexandratos, S.; Schleyer, P. v. R.; Streitwieser, A. Jr.; Schaefer, H. F., III J. Am. Chem. Soc. 1978, 100, 5695.

7. (a) Dewar, M. J. S.; Rzepa, H. S. J. Am. Chem. Soc. 1978, 100, 777 (b) Dewar, M. J. S.; Rzepa, H. S. Inorg. Chem. 1979, 18, 602.
8. Bews, J. R.; Glidewell, C. J. Organomet. Chem. 1981, 219, 279. See also, Glidewell, C. J. Organomet. Chem. 1981, 217, 273.
9. Marynick, D. S. J. Am. Chem. Soc. 1981, 103, 1328.
10. Jutzi, P.; Kohl, F.; Hofmann, P.; Krüger, C.; Tsay, Y.-H. Chem. Ber. 1980, 113, 757.
11. Anh, N. T.; Elian, M.; Hoffmann, R. J. Am. Chem. Soc. 1978, 100, 110.
12. Dewar, M. J. S.; Thiel, W. J. Am. Chem. Soc. 1977, 99, 4899. For a discussion of the validity of the MNDO method, see Dewar, M. J. S.; McKee, M. L. Inorg. Chem. 1978, 17, 1075.
13. Rudolph, R. W. Acc. Chem. Res. 1976, 9, 446.
14. (a) Wade, K. J. Chem. Soc. Chem. Commun. 1971, 792 (b) Wade, K. Adv. Inorg. Chem. and Radiochem. 1976, 18, 1.
15. (a) Dewar, M. J. S.; Thiel, W. J. Am. Chem. Soc. 1977, 99, 4907 (b) Dewar, M. J. S.; McKee, M. L. ibid. 1977, 99, 5231 (c) Dewar, M. J. S.; Rzepa, H. S. J. Am. Chem. Soc. 1978, 100, 58 (d) Davis, L. P.; Guidry, R. M.; Williams, J. R.; Dewar, M. J. S.; Rzepa, H. S. Journal of Computational Chemistry 1981, 2, 433.
16. Cotton, F. A. "Chemical Applications of Group Theory" 2nd. Ed. Wiley-Interscience 1971.
17. Frasson, E.; Menegus, F.; Panattoni, C. Nature 1963, 199, 1087,
18. (a) Shibata, S.; Bartell, L. S.; Gavin, R. M., Jr. J. Chem. Phys. 1964, 41, 717 (b) Tyler, J. K.; Cox, A. P.; Sheridan, J. Nature 1959, 183, 1182.
19. Ewig, C. S.; Osman, R.; Van Wazer, J. R. J. Am. Chem. Soc. 1978, 100, 5017.
20. (a) Evans, S., D. Phil. Thesis Oxford University, 1972 (b) Egdell, R. G.; Fragala, I.; Orchard, A. F. J. Electron Spectrosc. and Relat. Phenom. 1978, 14, 467 (c) Cradock, S. Duncan, W. J. Chem. Soc. Faraday Trans. 2 1978, 74, 194.
21. Cowley, A. H.; Lattman, M. To be published.
22. Jutzi, P.; Seufert, A. Angew. Chem. Int. Ed. Engl. 1977, 16, 330.
23. Bartke, T.; Bjørseth, A.; Haaland, A.; Marstokk, K. M.; Møllendal, H. J. Organomet. Chem. 1975, 85, 271.
24. Contreras, J. G.; Tuck, D. G. Inorg. Chem. 1973, 12, 2596.
25. Gropen, O.; Haaland, A. J. Organomet. Chem. 1975, 92, 157.
26. Grundke, H.; Paetzold, P. I. Chem. Ber. 1971, 104, 1136.
27. Drew, D. A.; Haaland, A. Acta Chem. Scand. 1973, 27, 3735.
28. Stadelhofer, J.; Weidlein, J.; Fischer, P.; Haaland, A. J. Organomet. Chem. 1976, 116, 55.
29. Mertz, K.; Zettler, F.; Hausen, H. D.; Weidlein, J. J. Organomet. Chem. 1976, 122, 159.
30. Beclé, B.; Corfield, P. W. R.; Oliver, J. P. Inorg. Chem. 1982, 21, 458.

31. Schonberg, P. R.; Paine, R. T.; Campana, C. J. Am. Chem. Soc. 1979, 101, 7726.
32. Schonberg, P. R.; Paine, R. T.; Campana, C. F.; Duesler, E. N. Organometallics 1982, 1, 799.
33. Einstein, F. W. B.; Gilbert, M. M.; Tuck, D. G. Inorg. Chem. 1972, 11, 2832.
34. Jutzi, P.; Kohl, F.; Krüger, C.; Wolmershäuser, G.; Hofmann, P.; Stauffert, P. Angew. Chem. Int. Ed. Engl. 1982, 21, 70.
35. (a) Baxter, S. G.; Cowley, A. H.; Mehrotra, S. K. J. Am. Chem. Soc. 1981, 103, 5572 (b) Cowley, A. H.; Mehrotra, S. K. J. Am. Chem. Soc. 1983, 105, 2074.
36. See, for example, (a) Soluki, B.; Rosmus, P.; Bock, H.; Maier, G. Angew. Chem. Int. Ed. Engl. 1980, 19, 51 (b) Maier, G.; Mihm, G.; Reisenauer, H. P. Angew. Chem. Int. Ed. Engl. 1980, 19, 52.
37. For a review, see Ashe, A. J., III Acc. Chem. Res. 1978, 11, 153.
38. Cuthbertson, A. F.; Glidewell, C. J. Organomet. Chem. 1981, 221, 19.
39. Cowley, A. H.; Ebsworth, E. A. V.; Mehrotra, S. K.; Rankin, D. W. H.; Walkinshaw, M. D. J. Chem. Soc. Chem. Commun. 1982, 1099.
40. Davison, A.; Rakita, P. E. Inorg. Chem. 1970, 9, 289.
41. (a) Rossi, A. R.; Hoffmann, R. Inorg. Chem. 1975, 14, 365 (b) Elian, M.; Maynard, M. L.; Chen, D.; Mingos, D. M. P.; Hoffmann, R. Inorg. Chem. 1976, 15, 1148.
42. Maier, G.; Pfriem, S.; Schaefer, U.; Matusch, R. Angew. Chem. Int. Ed. Engl. 1978, 17, 520.
43. (a) Stohrer, W. D.; Hoffmann, R. J. Am. Chem. Soc. 1972, 94, 1661 (b) Kollmar, H.; Smith, H. O.; Schleyer, P. v. R., J. Am. Chem. Soc. 1973, 95, 5834 (c) Dewar, M. J. S.; Haddon, R. C. J. Am. Chem. Soc. 1973, 95, 5836 (d) Hehre, W. J.; Schleyer, P. v. R. J. Am. Chem. Soc. 1973, 95, 5837.
44. (a) Masamune, S.; Sakai, M.; Ona, H.; Jones, A. J. J. Am. Chem. Soc. 1972, 94, 8956 (b) Masamune, S.; Sakai, M.; Kemp-Jones, A. V.; Ona, H.; Venot, A.; Nakashima, T. Angew. Chem. Int. Ed. Engl. 1973, 12, 769.
45. (a) Luth, H.; Amma, E. L. J. Am. Chem. Soc. 1969, 91, 7515 (b) Weininger, M. S.; Rodesiler, P. E.; Gash, A. G.; Amma, E. L. J. Am. Chem. Soc. 1972, 94, 2135 (c) Rodesiler, P. F.; Auel, Th.; Amma, E. L. J. Am. Chem. Soc. 1975, 97, 7405.

RECEIVED June 16, 1983

8

Novel Cluster Interactions in Metalloboranes

NORMAN N. GREENWOOD

University of Leeds, Department of Inorganic and Structural Chemistry, Leeds LS2 9JT, England

Metal atoms have fewer valence electrons than orbitals available for bonding and in this they resemble boron. The consequences of this idea are examined and it is shown that many metals with electronegativities in the range 1.6-2.4 (B = 2.0) can subrogate boron atoms as vertices in polyhedral clusters. Such metalloboranes are often much more stable than the parent boranes or borane anions. Not only can metals mimic boron in known cluster geometries but the flexibility thus introduced can lead to novel and previously unsuspected cluster geometries. The construction of macropolyhedral clusters containing 17-20 vertices is also described.

Ralph Rudolph made major contributions to our understanding of the structure and bonding of polyhedral cluster compounds and he had an abiding interest in developing a rationale which would enable the structure of individual compounds to be systematized and related to each other. He independently arrived at a method of counting skeletal electrons which is now generally referred to as Wade's Rules, and this has had a decisive influence on our general perceptions of polyhedral cluster compounds. Related to this was his preoccupation with the problem of heteroatoms such as sulfur, and the number of electrons which such atoms contribute to the heteroborane clusters.

You will recall that boranes are now classified into various series as summarized in Table I. The formulae and structures can be rationalized on the basis of the number of skeletal electrons available for bonding and each B atom is considered to contribute 2 electrons in addition to the one used to form a terminal $B\text{-}H_t$ bond. Supernumerary H atoms form $B\text{-}H_\mu\text{-}B$ bridges or comprise the *endo*-H atom in a BH_2 group: BH_tH_{endo}; they occur in the "open face" of structures from which BH_t groups are notionally missing

0097–6156/83/0232–0125 $06.00/0

Table I. Borane Structures - Wade's Rules

Series	Parent Formula	Skeletal Electrons	Cluster Geometry (deltahedron = closed triangulated polyhedron)
precloso-	B_nH_n	2n	(n-1)vertexed deltahedron plus 1 capping BH
closo-	B_nH_{n+2}	2n+2	n-vertexed deltahedron
nido-	B_nH_{n+4}	2n+4	(n+1)deltahedron (n occupied)
arachno-	B_nH_{n+6}	2n+6	(n+2)deltahedron (n occupied)
hypho-	B_nH_{n+8}	2n+8	(n+3)deltahedron (n occupied)
conjuncto-	B_nH_m	---	two or more clusters joined by B-B or BBB bonds, or by sharing common vertices, edges, or faces

from the parent closed triangulated polyhedral structure in which all vertices are occupied. In general the structures reflect the fact that boron has one less valence electron than atomic orbitals available for bonding. Most metals also have fewer valence electrons than orbitals available for bonding and some years ago we initiated a research program to delineate the extent to which metals could act as polyhedral vertices in metallo-borane complexes. It turns out that virtually all metals except the most electropositive ones in Groups I-III of the Periodic Table can act in this way as honorary boron atoms. Indeed, the metalloboranes are frequently very much more stable than the corresponding parent boranes and borane anions.

Three examples from our own recent work will suffice to show how coordinated metal atoms can subrogate BH units in *closo-*, *nido-* and *arachno-*decaborane. Thus, the bright red nickela-*closo*-decaborane *closo*-[$(PMe_2Ph)_2(1\text{-}NiB_9H_7Cl_2\text{-}2,4)$] has been synthesized (1) in which the B_9 moiety is η^4-bonded to a capping L_2Ni group (Figure 1). The structural relationship to *closo*-$B_{10}H_{10}^{2-}$ is obvious and, as the metal centre formally replaces a BH^{2-} group which contributes 4e to the cluster, it can be considered as Ni^{IV}. We find, in general, that polyhapto borane ligands are very good at stabilizing formal high oxidation states of metals. Next we can mention a *nido*-analogue, *nido*-[$H(PPh_3)_2(6\text{-}IrB_9H_{13})$] (Figure 2): this stable, yellow iridadecaborane bears an obvious structural resemblance to *nido*-$B_{10}H_{14}$ itself and we have made several other closely related iridium and rhodium analogues (2). The compound was originally made in 1% yield by reacting *arachno*-$B_9H_{14}^-$ with Vaska's compound *trans*-[$Ir(CO)Cl(PPh_3)_2$] but, as will be seen later, the yield can be dramatically improved to >85% by the

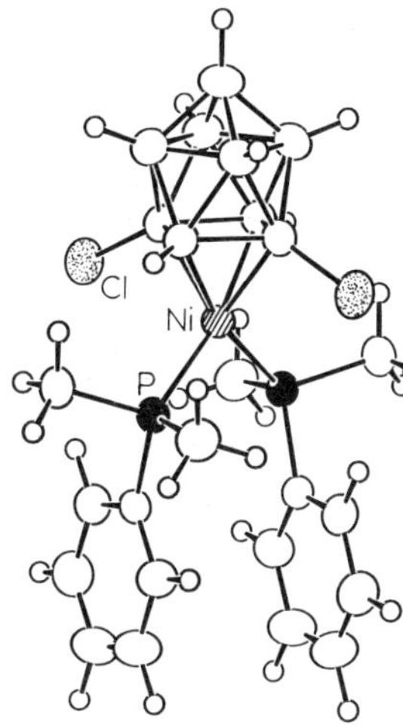

Figure 1. The structure of closo-*[$(PMe_2Ph)_2(1\text{-}NiB_9H_7Cl_2\text{-}2,4)$] showing the bicapped-square antiprismatic arrangement of the {NiB_9} cluster, which is analogous to* closo-*$B_{10}H_{10}^{2-}$ (1).*

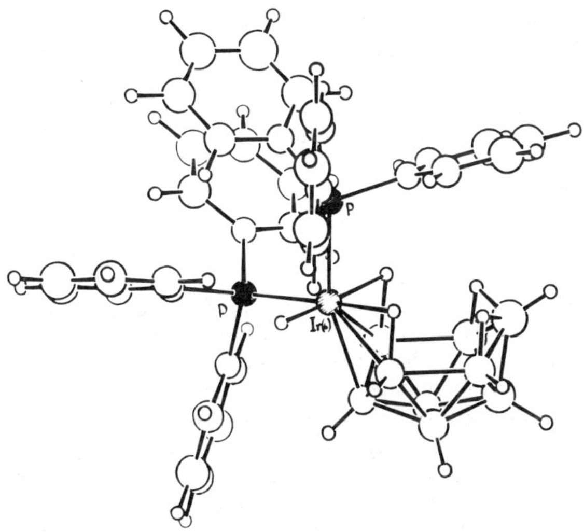

Figure 2. The structure of nido-*[$H(PPh_3)_2(6\text{-}IrB_9H_{13})$] showing the close similarity of* nido-*$B_{10}H_{14}$ (2).*

simple expendient of using the tris(tertiary phosphine) $[IrCl(PPh_3)_3]$ (3). Finally, the extremely stable diplatina-*arachno*-decaborane, $[(PMe_2Ph)_4(6,9\text{-}Pt_2B_8H_{10})]$, (Figure 3) can be readily prepared in good yield by a straight-forward three-step synthesis (4):

$$2B_9H_{14}^- \;+\; cis\text{-}[PtCl_2(PMe_2Ph)_2]$$

$$\downarrow\ CH_2Cl_2\ (12\ h/r.t.)$$

$$[Pt(B_8H_{12})(PMe_2Ph)_2]\ \text{(very pale yellow, 85\% yield)}$$

$$\downarrow\ 2KH/thf\text{-}CH_2Cl_2\ (4\ h/r.t.)$$

$$[Pt(B_8H_{10})(PMe_2Ph)_2]^{2-}\ \text{(not isolated)}$$

$$\downarrow\ cis\text{-}[PtCl_2(PMe_2Ph)_2]\ (12\ h)$$

$$[\{Pt(PMe_2Ph)_2\}_2(B_8H_{10})]\ \text{(colourless, 55\% yield)}$$

This was the first example of such a dimetalladecaborane and its structural similarity to derivatives of *arachno*-$B_{10}H_{14}^{2-}$ is clear. The examples in Figures 1-3 will suffice to illustrate the subrogation of B by M in boron hydrides and similar examples could now be given with over 30 different metals and virtually all boranes from B_1 to B_{12} (5).

But an even more exciting possibility can now be envisaged. Instead of merely mimicing the known borane structures we can attempt to use the flexibility of metal atoms to construct polyhedral clusters of geometry as yet unknown among the parent boron hydrides or, indeed, by using the higher connectivities and oxidation states of some metals, to devise metalloborane cluster geometries which would be incapable of synthesis if boron alone was the polyhedral vertex atom. We have found that such structures can indeed be synthesized and so metals now graduate from being "honorary boron atoms" to being "flexi-boron atoms". Two approaches are possible. The first is to modify reaction conditions and reagents systematically so that unpredicted and even unpredictable compounds emerge. We can then use these new perceptions in the second approach, which is the purposeful design of reactions which are specifically aimed at incorporating new structural features based on the properties of the metal centre used. Both approaches have their place in synthetic chemistry: the first is tremendously exciting, *when* it works; the second is tremendously satisfying, *if* it works.

As an example of the first method we can enquire what would happen if the *arachno*-$B_9H_{14}^-$ used in the last preparation above were replaced by the isoelectronic neutral species *arachno*-$[4\text{-}(Me_2S)\text{-}7\text{-}(MeO)B_9H_{12}]$. In fact, the predominant product (50% yield) is the expected methoxy derivative *arachno*-$[(PMe_2Ph)_2(Pt\text{-}B_8H_{11}OMe)]$ plus a smaller amount of the known *arachno*-$[PMe_2Ph)_2$-

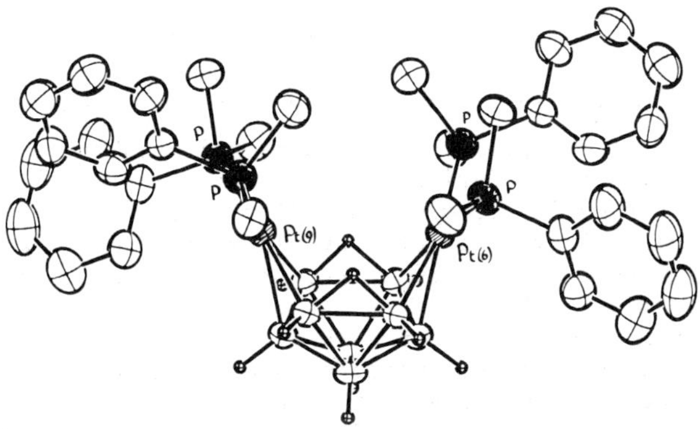

Figure 3. The structure of arachno-$[(PMe_2Ph)_4(6,9\text{-}Pt_2B_8H_{10})]$ *showing its structural similarity to* arachno-$B_{10}H_4^{2-}$ (4).

$(PtB_3H_7)]$. In addition, however, the products included a 3% yield of the diplatina cluster *conjuncto*-$[(PMe_2Ph)_2(Pt_2B_8H_{14})]$ shown in Figure 4a, (6). This structure is unparalleled in borane chemistry or in any other known compound; it features a central linear P-Pt-Pt-P coordinated by an η^3-B_6H_9 group (structurally derived from *arachno*-B_6H_{10}) and a bridging B_3H_5 (Figure 4b). The Pt-Pt distance is 262.1 pm and the four atoms in the central diamond Pt(1)B(2)Pt(2)B(7) are essentially coplanar (dihedral angle 0.9°). The structure is therefore entirely different from that of the formally isoelectronic *arachno*-diplatinadecaborane cluster in Figure 3.

Another product from the reaction of *cis*-$[PtCl_2(PMe_2Ph)_2]$ with *arachno*-$[4\text{-}(Me_2S)\text{-}7\text{-}(MeO)B_9H_{12}]$ is *conjuncto*-$[(PMe_2Ph)_2$-$(Pt_2B_{12}H_{18})]$ which was isolated in 35% yield. This compound, which had previously been isolated in our laboratory from the reactions of *cis*-$[PtCl_2(PMe_2Ph)_2]$ with 6,6'-(*nido*-$B_{10}H_{13})_2O$ (7), and with *arachno*-$B_9H_{14}^-$ (4), is particularly stable and features two η^3-B_6H_9 groups centrosymmetrically coordinating a linear P-Pt-Pt-P group as shown in Figure 5a. The structural similarities with the cluster in Figure 4 are striking, the main differences being in the orientation of the two phenyl groups of the tertiary phosphine ligands, and in the removal of 4 further B atoms from the upper B_6 subcluster in Figure 5 to give the B_2 group in Figure 4. The detailed reaction path by which the Pt-Pt bonds are formed and the initial B_9 (or B_{10}) cluster are degraded to B_6 and B_2 clusters has not yet been elucidated; it is a fascinating problem, the resolution of which will undoubtedly lead to more efficient syntheses of these intriguing compounds.

Another classic method of obtaining higher boranes from lower boranes is thermolysis. Accordingly, we studied the effect of heat on appropriate metallaboranes and found that in some cases we obtained metal-mediated oligomerization to give novel macropolyhedral clusters. For example, when the pale yellow compound $[Pt(B_8H_{12})(PMe_2Ph)_2]$ (see above) was heated in boiling toluene (111°C) several products were obtained including a 2% yield of the flame-red, air-stable, compound $[(PMe_2Ph)\{PtB_{16}H_{18}$-$(PMe_2Ph)\}]$ m.p. 95° (8). The completely unprecedented structure, illustrated in Figure 6a, can be viewed as an edge-fused *conjuncto*-borane consisting of a *nido*-$\{PtB_7\}$ cluster and a *nido*-$\{PtB_{10}\}$ cluster fused at a common Pt-B edge. One approach to a description of the bonding is indicated in Figure 6b - the borane ligand is a phosphine substituted 16-vertex cluster formally derived from the as yet unknown *conjuncto*-$B_{16}H_{24}$. This part of the structure clearly comprises a *nido*-B_6 subcluster bonded to the B(6) atom of a *nido*-B_{10} subcluster in a way very reminiscent of that just reported at this Meeting for the B_6 plus *nido*-B_9 structure of *conjuncto*-$B_{15}H_{23}$ (9). The implications of this for the purposeful synthesis of other macropolyhedral *conjuncto*-boranes is considerable.

We have also had some exciting indications of metal-

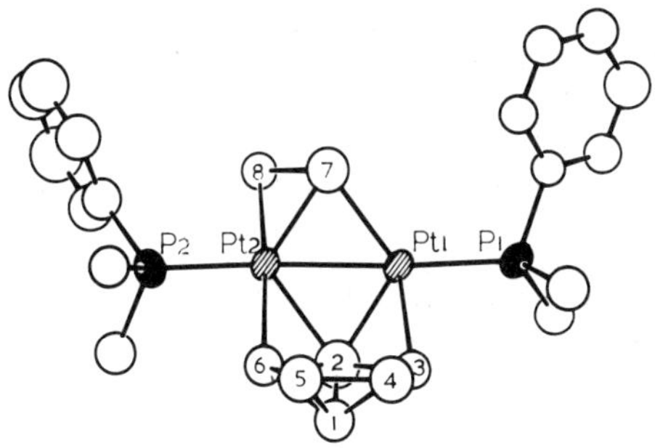

Figure 4a. The structure of conjuncto-*[$(PMe_2Ph)_2(Pt_2B_8H_{14})$] showing the central linear P–Pt–Pt–P group coordinated by an {η^3-B_6H_9} group and a bridging {B_3H_5}.*

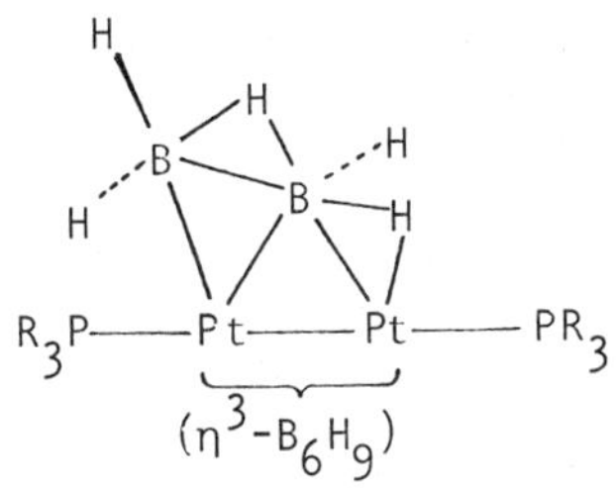

Figure 4b. The disposition of the 5 H atoms in {B_3H_5} as revealed by NMR spectroscopy (6).

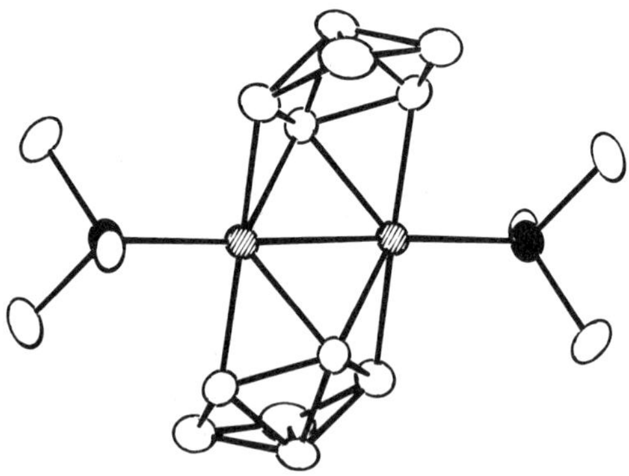

Figure 5a. The cluster geometry of conjuncto-*[$(PMe_2Ph)_2Pt_2(\eta^3$-$B_6H_9)_2$]; only the* ipso-*C atoms of the two phenyl groups are shown and these are centrosymmetrically disposed above and below the plane of the paper.*

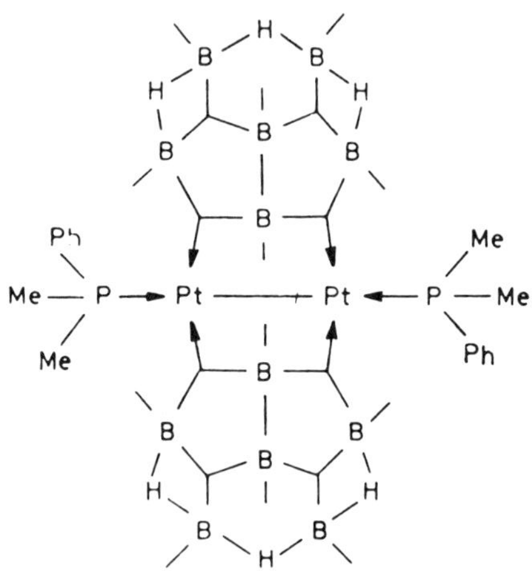

Figure 5b. A simple localized valence bond description of the conjuncto-*[$(PMe_2Ph)_2Pt_2(\eta^3$-$B_6H_9)_2$] structure (7).*

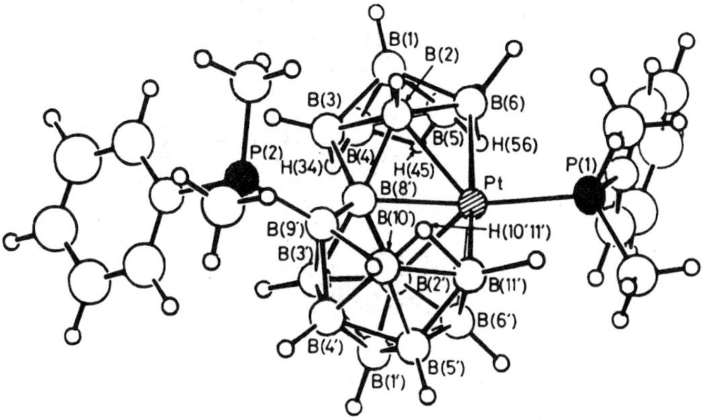

Figure 6a. The molecular structure of the 17-vertex conjuncto-*platinaborane* $[(PMe_2Ph)\{PtB_{16}H_{18}(PMe_2Ph)\}]$.

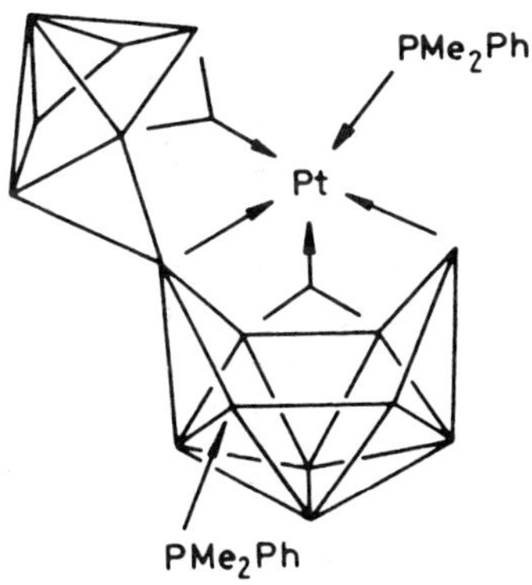

Figure 6b. An outline diagram of the macropolyhedral borane ligand geometry and its interaction with the Pt center (8).

mediated cluster construction in some work with the platinum complexes of $B_{18}H_{20}^{2-}$ (10). One of the products of the reaction between *cis*-[$PtCl_2(PMe_2Ph)_2$] and the centrosymmetric edge-fused *conjuncto-anti*-$B_{18}H_{22}$ in the presence of a strong deprotonating agent is the novel, green, *face-fused* diplatina *conjuncto*-borane [$(PMe_2Ph)_4\{Pt_2$-η^4,(η^4+η^2)-*anti*-$B_{18}H_{16}\}$] shown in Figure 7. A similar confacial triangular feature can also be constructed by the mild thermolysis of another product of the initial reaction [$(PMe_2Ph)_2$(Pt-η^1-η^2-*anti*-$B_{18}H_{20}$)] (Figure 8a): this undergoes an astonishing rearrangement with loss of H_2 to give the *face-fused* [(PMe_2Ph)(Pt -η^4-*anti*-$B_{18}H_{18}$)] (11) as shown in Figure 8b. Here the presence of a Pt vertex has facilitated a cluster reorganization which certainly does not occur in its absence and which augers well for the use of metalloborane intermediates to synthesize new macropolyhedral clusters with novel geometrical features.

A more subtle variation of the theme would be to synthesize structural isomers which differ in some specific way from the geometry of known borane clusters. One such example is implicit in the diplatina structure shown in Figure 5a which is clearly the *transoid* variant of the known *cisoid* structure of the "isoelectronic" *conjuncto*-$B_{14}H_{20}$ (12). Another structural variant would be the position of the vacant site in going from a *closo*- to a *nido*-cluster. For example, the cluster geometry of the normal *nido*-$B_{10}H_{14}$ and its metallo derivatives can be envisaged as being formed by removal of the unique, highly-connected {BH} vertex from $B_{11}H_{11}^{2-}$ to give a nest-like cluster as illustrated in Figure 2. An alternative would be to retain the highly-connected vertex as a metal centre and remove one of the adjacent {BH} vertices to give an *iso-nido*-metalladecaborane. Just such a cluster was found among the minor products of the reaction of *arachno*-$B_9H_{14}^-$ with *trans*-[$Ir(CO)Cl(PPh_3)_2$]: as already mentioned the major product was *nido*-[$H(PPh_3)_2$(6-IrB_9H_{13})], (Figure 2) but there were also trace amounts of a pale violet crystalline complex which was shown by X-ray studies (13) to be *iso-nido*-[$(PPh_3)_2$(7-$IrB_9H_{10}\cdot PPh_3$)] (Figure 9). The distances from Ir(7) to B(2,3,4,8,10) are 236, 217, 236, 233, and 233 pm respectively and the distance B(8)-B(10) across the "open" face is 222 pm, suggesting some residual interactions as represented by the open connector in the Figure. All other nearest-neighbour B-B distances were within the normal range of 167-187 pm. There are terminal H atoms (not shown) on each boron except B(9), which carries a pendant PPh_3 ligand and the two remaining H atoms are associated with the open face Ir(7)B(8)B(9)B(10) - they are inequivalent in solution at low temperature but undergo mutual exchange with $\Delta G^{\ddagger}$ ~45 kJ mol at −10°C. Indeed, the cluster seems to be in some sense intermediate between a *nido*- and a *closo*-geometry with Ir-H_{μ}-B bridge hydrogen atoms interconverting with terminal Ir-H_t.

As a final example of the construction of a modified cluster

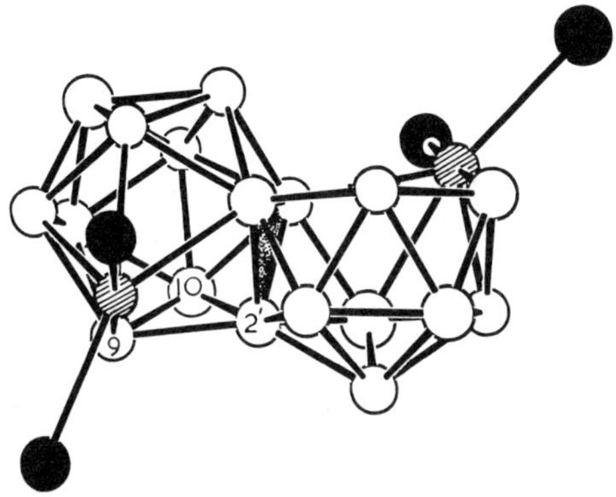

Figure 7. The cluster geometry of the green diplatina product conjuncto-*[(PMe$_2$Ph)$_4${Pt$_2$-η^4, (η^4 + η^2)*-anti-*B$_{18}$H$_{16}$}] showing the confacial fusion of the two subclusters across the shaded triangle* (10).

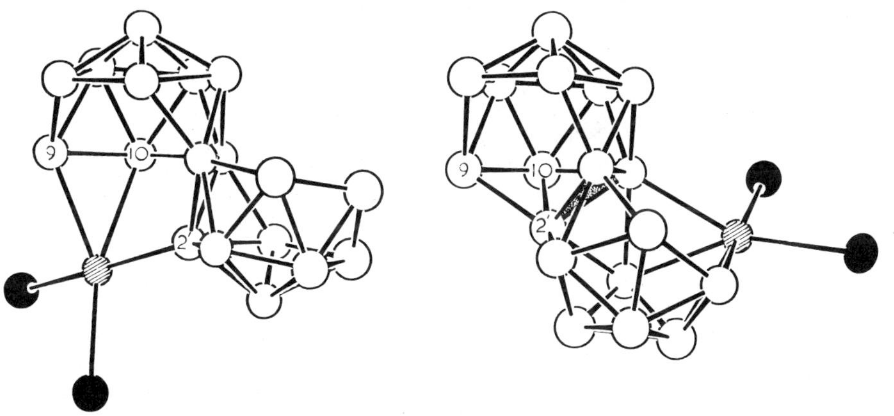

Figure 8. Left, the cluster geometry of iso-*[(PMe$_2$Ph)$_2$(μ-Pt-η^1, η^2*-anti-*B$_{18}$H$_{20}$)]; during thermolytic rearrangement a new bond is formed between B(9), B(10), and B(2'), and the bridging {Pt(PMe$_2$Ph)$_2$} group appears to migrate to the open face of the right-hand subcluster. Right, the cluster geometry of the thermolysis product [(PMe$_2$Ph)$_2$(Pt-η^4-*anti-*B$_{18}$H$_{20}$)] showing the new confacial triangular linkage (shaded) between the two subclusters* (11).

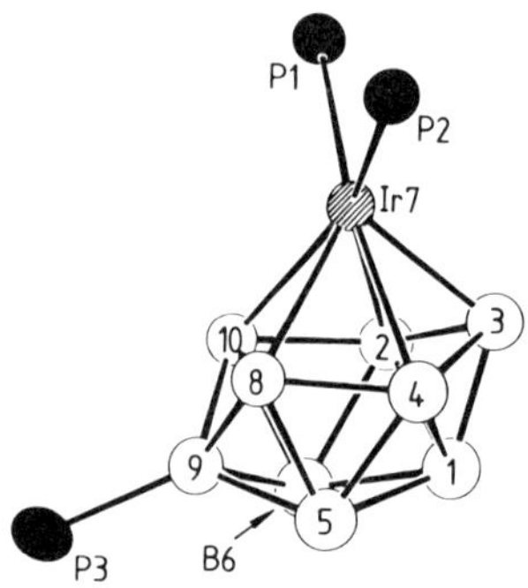

Figure 9. Cluster geometry of iso-nido-*[$(PPh_3)_2(7\text{-}IrB_9H_{10} \cdot PPh_3)$]* (13).

geometry we can mention the astonishingly facile dehydrogenation of *nido*-[$H(PPh_3)_2(6\text{-}IrB_9I_{13})$] (Figure 2): mild thermolysis of this compound at 85°C results in the quantitative elimination of $3H_2$ to give the bright orange-yellow *iso-closo*-[$H(PPh_3)(PPh_2C_6H_4)(1\text{-}IrB_9H_8)$] (3). As seen in Figure 10 this complex features a ten-vertex *closo* structure with idealized C_{3v} symmetry rather than the more usual 4-fold symmetry of $B_{10}H_{10}^{2-}$ and its derivatives (Figure 1).

The distances from Ir(1) to B(3,5,7) are 216, 219 and 215 pm respectively and to the adjacent B(2,4,6) are 246, 239, and 238 pm. The Ir-H_t distance is ∿157 pm. The local C_{3v} symmetry of the $IrB_3B_3B_3$ cluster has not previously been encountered in metalloborane chemistry though a similar disposition of vertices has been observed in, for example, the dimetalladicarbaborane [$1,6\text{-}(\eta^5\text{-}C_5H_5)_2\text{-}1,6,2,3\text{-}Fe_2C_2B_6H_8$] (14). A formal electron count for the compound in Figure 10 requires 22 skeletal electrons (2n+2) for the *closo* 10-vertex cluster bonding: of these, 18 are supplied by the 9 B atoms, leaving 4 to be contributed by the Ir atom. This, together with the Ir-H_t bond indicates that the compound can be regarded as an Ir^V complex in which the high oxidation state is stabilized by coordination to the polyhedral borane ligand. Other examples of Ir^V borane complexes are also known (15,16). A further interesting aspect of the molecular structure in Figure 10 is the incidence of *ortho*-cycloboronation at B(2). This is now recognized as a significant feature of several cluster reorganization reactions and has important mechanistic implications which have recently been reviewed (17).

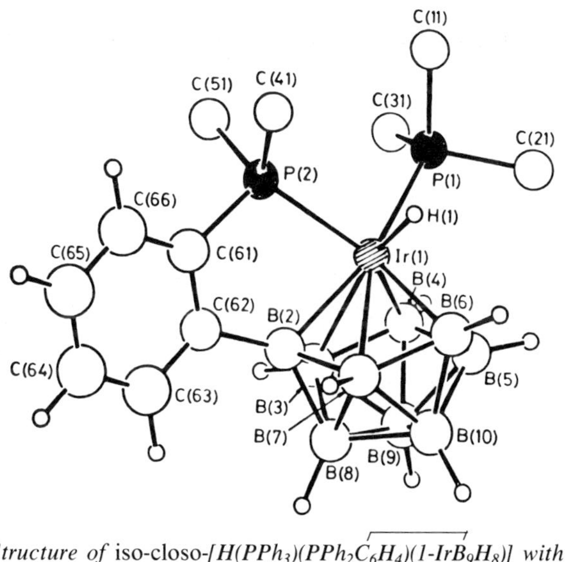

Figure 10. Structure of iso-closo-*[$H(PPh_3)(PPh_2C_6H_4)(1\text{-}IrB_9H_8)$] with* P-*phenyl groups omitted for clarity, except for their* ipso-*C atoms; all H atoms were located (3).*

Literature Cited

1. Greenwood, N.N.; Hails, M.J.; Kennedy, J.D.; McDonald, W.S. J. Chem. Soc. Dalton Trans. 1983, to be published.
2. Boocock, S.K.; Bould, J.; Greenwood, N.N.; Kennedy, J.D.; McDonald, W.S. J. Chem. Soc. Dalton Trans. 1982, 713.
3. Bould, J.; Greenwood, N.N.; Kennedy, J.D.; McDonald, W.S.; J. Chem. Soc. Chem. Commun., 1982, 465.
4. Boocock, S.K.; Greenwood, N.N.; Hails, M.J.; Kennedy, J.D.; McDonald, W.S. J. Chem. Soc. Dalton Trans. 1981, 1415.
5. Greenwood, N.N.; Kennedy, J.D. "Metal Interactions with Boron Clusters"; Grimes, R.N., Ed.; Plenum Press, New York, 1982, Chapter 2.
6. Ahmad, R.; Crook, J.E.; Greenwood, N.N.; Kennedy, J.D.; McDonald, W.S. J. Chem. Soc. Chem. Commun., 1982, 1019.
7. Greenwood, N.N.; Hails, M.J.; Kennedy, J.D.; McDonald, W.S. J. Chem. Soc. Chem. Commun. 1980, 37.
8. Beckett, M.A.; Crook, J.E.; Greenwood, N.N.; Kennedy, J.D.; McDonald, W.S. J. Chem. Soc. Chem. Commun., 1982, 552.
9. Huffman, J.C.; Ragatz, P.; Schaeffer, R.; Abstracts of Papers, 184th ACS National Meeting, Kansas City, Missouri, Sept. 12-17, 1982, INOR 58.
10. Cheek, Y.M.; Greenwood, N.N.; Kennedy, J.D.; McDonald, W.S. J. Chem. Soc. Chem. Commun., 1982, 80.
11. Cheek, Y.M.; Crook, J.E.; Greenwood, N.N.; Kennedy, J.D. unpublished results 1982.
12. Hufmann, J.C.; Moody, D.C.; Schaeffer, R. J. Am. Chem. Soc. 1975, 97, 1621; Inorg. Chem. 1981, 20, 741.
13. Bould, J.; Greenwood, N.N.; Kennedy, J.D.; McDonald, W.S. unpublished work 1982.
14. Callahan, K.P.; Evans, W.J.; Lo, F.Y.; Strouse, C.E.; Hawthorne, M.F., J. Am. Chem. Soc. 1975, 97, 296.
15. Crook, J.E.; Greenwood, N.N.; Kennedy, J.D.; McDonald, W.S. J. Chem. Soc. Chem. Commun. 1981, 933.
16. Bould, J.; Crook, J.E.; Greenwood, N.N.; Kennedy, J.D.; McDonald, W.S. J. Chem. Soc. Chem. Commun., 1982, 346.
17. Greenwood, N.N. in Chisholm, M.H. (ed.) Inorganic Chemistry Toward the 21st Century, Bloomington, Indiana, May 16-19, 1982, ACS Symposium Series, in press.

RECEIVED March 30, 1983

9

Molecular State Fingerprints and Semiempirical Hypersurface Calculations

Useful Correlations to Track Short-Lived Molecules

H. BOCK

University of Frankfurt, Chemistry Department, D-6000 Frankfurt/Main 50, Niederurseler Hang, West Germany and University of Michigan, Chemistry Department, Ann Arbor, MI 48109

R. DAMMEL and B. ROTH

University of Frankfurt, Chemistry Department, D-6000 Frankfurt/Main 50, Niederurseler Hang, West Germany

Nowadays even the synthetic chemist, relying on one of the many well-tested semiempirical quantum chemical procedures, can perform approximate hypersurface calculations for his somewhat larger molecules. Despite some pitfalls - for example, concerning reaction pathways without the molecular dynamics involved - valuable suggestions are provided on the energy differences and the structural changes between individual molecular states, e.g. in redox reactions. As an advantage for the experimentalist, the correlation of theoretical predictions with easily accessible measurement data such as ionization patterns or spin coupling multiplets is strongly emphasized: thus, not only the validity of underlying assumptions can be checked but also the numerical precision added to the calculated trends.

For the practical design of hypersurfaces, i.e. cuts through the (3n-6)dimensional hyperspace, some hints are outlined. The main purpose, however, is to illustrate the usefulness of hypersurface calculations especially for the detection, identification and characterization of unstable molecules. Examples chosen comprise the structure of RS-C≡C-SR, the relative stability of thioacroleine isomers C_3H_4S, the structural changes accompanying the oxidation of hydrazine and some of its derivatives, the isomerization of tetrahedrane to cyclobutadiene both thermally as well as on oxidation, the predicted existence of F_2SS and nonexistence of Cl_2SS or H_2SS, and, finally, some aspects of the thermal decomposition of methyl and vinyl azides.

The 'paradigm for electronic requirements of clusters' as emphasized by the late R.W. Rudolph - to whom this memorial symposium of the American Chemical Society is dedicated - and also by

0097-6156/83/0232-0139 $07.75/0

K. Wade, R.E. Williams and others(1) is one of the many intrinsic relationships between the topological structure, the charge distribution and the minimum potential energy of molecules in their ground states(5). Together with numerous other theoretically derived rules or empirically found regularities - starting with early contributions such as the (4n+2) π-electron postulate of E. Hückel, the orbital diagrams of A.D. Walsh or the Gillespie-Nyholm electron pair repulsion model, to mention just a few (5) -, these useful approaches are more and more applied by chemists in their daily work to stimulate research and to rationalize results. And it is within this scope and on this occasion, that experience with semiempirical hypersurface calculations gathered over several years will be summarized without diving into the theoretical foundations repeatedly reviewed by others elsewhere, e.g. (6,7).

Molecules change their properties on acquisition or loss of energy and the individual state of a molecule can be characterized by the energy difference to a preceding or to a subsequent state, as well as by the respective charge distribution. The 'molecular state' picture has been and is used by the chemist for his own benefit in many ways, for example, to disentangle the multitude of known compounds and their numerous properties by defining 'parent systems' and 'substituent perturbations', or more generally, to compare 'equivalent states' of 'chemically related', preferably iso(valence)electronic molecules (5). The wealth of data resulting from physical measurements of molecular properties has been largely rationalized in this way.

A special case are those molecules which are too unstable to survive storage under normal laboratory conditions. Usually, they are detected, identified and characterized by their molecular state 'fingerprints'. Among the multitude of methods available from the physical armory, photoelectron spectroscopic ionization patterns(4) and electron spin resonance signal multiplets (3) may serve as examples for well-developed tools, which are easily accessible to the preparative chemist and which provide information on either the energy differences between or the structure in individual molecular states(5). Quite often, however, the short-lived intermediates produced are unique species lacking analogy to already known compounds. If their size does not exceed a few atoms, rigorous quantum chemical calculations including most of the correlation energy frequently are well-suited to reproduce the measured data with a numerical accuracy sufficient to unambiguously identify the compound prepared(5). On the other hand, for larger molecules, especially those of low symmetry, a large gap opens between the reliability of the quantum chemical predictions and the computing time needed for them(6,7). Fast semiempirical calculation procedures are presently one reasonable compromise - expecially if applied to generate molecular property 'hypersurface' maps providing insight into the overall behaviour of the ensemble of atoms investigated, and if supported by as close as possible correlation with experimental data.

The approach outlined above i.e. substituting more precise quantum chemical calculations at a given fixed geometry by a screening process varying structural parameters and improving this approximation by fit to measurement, however, contains numerous inherent pitfalls. Most evident is the discrepancy between the reality that the energy within a three-dimensional molecule is distributed over its 3n-6 internal degrees of freedom and the rather limited visualization on two-dimensional hypersurfaces with the third coordinate representing a molecular property e.g. its total energy. Even if those two-dimensional cuts through (3n-6)-dimensional hyperspaces are guided by the chemist's intuition and some simplifying recipes discussed later, only the correspondence to all molecular data measured independently will justify the underlying assumptions and severe restrictions involved. On the other hand, some otherwise inaccessible insight might compensate for the risk taken.

This applies especially to the pathway of chemical reactions: despite all that is known about reaction behaviour, understanding of it is rather shallow(8). Within this area of intensive research stretching from molecular beam experiments (7) to rigorous quantum mechanical calculations of reaction channels(7) the chemist successfully anchors himself to life-belts such as the well-tested Woodward-Hoffmann rules or e.g. the X-ray structure relationships by Bürgi and Dunitz(6). Short-lived molecules, which are by definition reactive, might also add some facets: starting from a given structure, i.e. without preceding collision and energy transfer between the deformed reactants, the products are formed. And despite all reservation concerning two-dimensional cuts through 3n-6 dimensions and the neglect of molecular dynamics (9), semiempirical hypersurface calculations can at least stimulate experiments (cf. Example VI given later).

In the following, experience on semiempirical hypersurface calculations - mostly applying the MNDO approximation(9) - is reported. As will be exemplified, correlation with experimental data on structures, isomer stabilities, ionization patterns or ESR coupling constants ranged between satisfactory and acceptable. Altogether, they have been of great help in tracking some short-lived molecules.

Some Hints on the Design of Hypersurface Calculations

Background Philosophy. Within the framework of the Born-Oppenheimer approximation (11), the solutions of the Schroedinger equation, $H\Psi = E\Psi$, introduce the concept of molecular structure and, thereby, the total energy hyperspace: provided that the electronic wave function varies only slowly with the nuclear coordinates, electronic energies can be calculated for sets of fixed nuclear positions. The total energies i.e. the sums of electronic energy and the energy due to the electrostatic repulsions, in return, form a potential in which the motions of the nuclei occur. Variation of some structural parameters then generates total energy hypersurfaces and, in an approximate way, the changes in the nuclear coordinates might be considered as moving atomic mass points in the corresponding potential.

For chemical purposes, substitution of total energy hypersurfaces by those based on the heat of formation seems more reasonable, with the difference given by the zero point energy corrections. However, their calculations from first principles can be rather cumbersome (12) and, moreover, for a given variation of some nuclear coordinates it usually can be assumed that the change in zero point energy is small compared to that of the total energy. On the other hand, several semiempirical quantum chemical procedures which are appropriately parametrized often yield satisfactory approximations for molecular heats of formation (10) and, therefore, ΔH_f hypersurfaces have become rather common.

The Selection of Hypersurface Coordinates. Energy hypersurfaces for nonlinear molecules comprising n atoms depend parametrically on all 3n-6 internal nuclear coordinates. Considering all of them, however, would require computing times which already for medium-sized molecules would easily exceed human life-span (6), and, obviously, lead to a visualization chaos. Own experience in the design of hypersurface calculations can be summarized in the following (trivial) hints:

i. Consultation of molecular structure data compilations can guide the chemical intuition, which parts of the molecule are most strongly affected by a given perturbation and which subunit structural parameter changes might be neglected within a preliminary screening.

ii. Transformation of Cartesian coordinates of individual nuclear positions into angular coordinates, whenever possible, is strongly advisable and will, in general, considerably simplify any structural optimization. Characterizing the molecule by bond distance, bond angles and especially by dihedral angles not only helps to define the essential perturbation of the system and to choose less affected subunits with fixed structural parameters, but also to emphasize the symmetry of the system more strongly.

iii. Symmetry restrictions, whenever justifiable, will also considerably reduce the computing time needed for the calculation of potential energy minimum structures, e.g. for optimizations using a center of inversion by almost one half. In addition, deviations from known equilibrium structures denoted by their symmetry can be avoided. Sometimes, definition of local symmetry for specific subunits is recommended. On the other hand, symmetry constraints outside potential energy minima should be handled with great care.

iv. Sometimes "dummy" centers have to be introduced, relative to which movements are defined, e.g. for 3 atoms in a row one outside the connecting line to avoid an additional degree of freedom due to under-determinated coordinates. Quite often, even for fully defined coordinate systems, the introduction of 'dummies' relative to which certain movements are carried out can be of advantage.

v. For many problems it can be very helpful to first estimate how many and which structural isomers have to be included. Graph-theoretical approaches for given coordination numbers of the individual centers allow one to generate the set of possible connectivity combinations for a certain ensemble (13), from which the most probable can be selected for preliminary pin-point calculations. The resulting total energies can be used to choose all structures to be considered within a certain, chemically feasible energy range.

vi. Last, but not least, the hypersurface coordinates - for all practical purposes not exceeding three - have to be chosen. With reference to more detailed discussions of the many difficulties involved (6), the obvious first approach is to try to connect as many of the precalculated (cf. v.) structures within the given energy range by as few as possible perturbation coordinates, preferably by angle variations (cf. ii. and iii.)

The application and the usefulness of the above hints as well as some more special ones will be illustrated by the Examples I to VI presented later on.

Computing Time Estimates. Due to the complexity of the problem - starting from the size of the molecule(s) incorporated and the calculation procedure chosen, via the number of parameters varied and the density of calculations performed along the perturbation coordinate(s), to the information wanted and, finally, the way the results are printed or plotted - there is no general approach to evaluate hypersurface computation costs. Nevertheless, some ratio between investment and product can be approximated as soon as the problem and the answers aimed at are specified.

For all of the examples presented (Figures 1 to 6), due to the close correlation with experimental data emphasized, only semiempirical calculations were needed, and especially the MNDO procedure (10) proved to be reliable for most purposes. For instance, Koopman's deviations (5), $\Delta IE_n^V = -\Delta\varepsilon_J^{MNDO}$, between photoelectron spectroscopically determined spacings in vertical ionization patterns, ΔIE_n^V, and differences in calculated MNDO eigenvalues, $-\Delta\varepsilon_J^{MNDO}$, hardly ever exceeded standard errors $SE \leq 0.3$eV in corresponding linear regressions. Also, MNDO optimized potential energy minimum geometries in most cases closely resemble known molecular structures. In addition to the obviously achieved reasonable parametrization for main group elements up to n = 3, i.e. including elements such as silicon, phosphorus or sulfur, the Davidson-Fletcher-Powell subroutine included in the MNDO program (10) to effectively reach potential energy surface minima by an energy gradient method, usually leads to rapid convergence of the calculation performed. To provide some idea on computing times needed, it takes about 2 minutes to calculate one point on the C_2H_3N energy surface (Figure 6) using a DEC 1091 computer.

More important with respect to computing costs, however, is the design of the hypersurface calculations as outlined in the preceding subchapter. Varying the particularly convenient sets of bond distances, bond angles and dihedral angles only for the most strongly perturbed subunits considerably lowers expenses. Most important, obviously, is the appropriate choice of 2 perturbation coordinates out of the 3n-6 possible ones. According to the permutation formalism, $\binom{3n-6}{2} = \frac{3}{2}(3n^2-13n+14)$ combinations can be chosen, e.g. for the six-atom ensemble C_2H_3N already 66. If all essential isomeric species within a chemically reasonable total energy range can be arranged on a surface specified by only one pair of coordinates, the question to answer next is how many points have to be calculated along each parameter axis to obtain the necessary resolution. For an n by m frame, the careful specification of the n·m computations can be quite essential, e.g. for the 240 hypersurface points calculated for the vinylnitrene rearrangement (Figure 6), already 240 x 2 = 480 minutes = 8 hours CPU time were needed - not counting the backward calculations needed to control the results obtained (6), and the effort to achieve appropriate plotting.

Returning to the general approach, a triatomic system has (only! cf. 7) 3 degrees of freedom, the C_2H_3N moiety already 9. A complete hypersurface calculation for the latter with an (insufficient) spacing of only 6 points along each coordinate would require 6^9 = 10.077.696 energy calculations. Switching from fast semiempirical procedures via ab initio calculations with unsaturated basis sets to rigorous quantum chemical treatment beyond the Hartree-Fock limit, which includes a high percentage of correlation energy (7) and also permits vibrational levels (7) to be incorporated with the accuracy needed, can easily extend the computing time needed for a single point by a factor 10^2 to 10^3.

It is this gap between ambition and reality, where especially for larger entities of low symmetry, chemical intuition and - above all - close correlation to experimental data moves in. And the very degree of abstraction, simplification and approximation involved already for molecules in their ground states or, more generally, in potential energy surface minima always has to be kept in mind when the advantages are advertized: predicting and identifying unique (short-lived) molecules, and (with great caution) rationalizing pathways for their generation and decomposition.

Some Remarks on Reaction Pathways. So far, hypersurface calculations with close correlation to (spectroscopic) molecular data have been considered. Outside the potential minima, the uncertainty of the calculated results increases considerably. Referring explicitly to the clear-cut and comprehensible treatment of problems and pitfalls involved in approximating reaction pathways and especially hypersurface saddle-points by K.E. Müller (6), a few remarks may suffice to outline the scope of additional difficulties encountered.

Any description of the rather complex channel of a reaction between molecules - heated up by collisions with e.g. the wall or other molecules, crashing into the reaction partner within a cone of successful approach, being deformed and participating in energy transfer within the reaction complex which then separates into the products - by defining reaction coordinates can only be justified by its conceptual simplicity and the visualizibility of some (essential) features. With the energy storage within a molecule over its 3n-6 degrees of freedom as repeatedly discussed, it remains to be pointed out that even a 'properly' calculated two-dimensional hypersurface suffers from several inherent uncertainties.

Calculating within the same approach forward and backward between 2 minima on a twodimensional potential energy surface may - depending on its shape - not yield an identical saddle-point (6). Rigorous determination of the minimum energy barrier is feasible by more advanced methods (6) such as the 'Linear Synchrons Transit' or the 'Ring Valley Point' procedures but requires special effort. Even if the saddle-point is correctly localized, the trace of the reaction path calculated within both attached minimum energy valleys often remains ambiguous: the wider the valley and the softer its flanks, the more the calculated trace can meander around the bottom line. Especially valleys which are bent in a way that the calculated trace becomes orthogonal to one of the coordinates i.e. independent of it, often leads to ill-defined pathways - a methodical shortcoming originating from the discussed restriction of the (3n-6)dimensional hyperspace to a twodimensional projection on a vertical energy plane. Other ambiguities may arise, when especially semiempirical calculations are numerically insufficient to differentiate between several reaction pathways of almost comparable energy.

Returning to the general background philosophy outlined and specifically to short-lived intermediates of medium size, semiempirical hypersurface calculations despite of their inherent short-comings may help the chemist in several aspects as a guideline and a rationale in his experiments. The molecular state patterns calculated for molecules and their ions can be improved considerably by a preceding localization of their total energy minimum, i.e. approximating their structure more closely. And, by comparison with measured data not only detection, identification and characterization of the species generated becomes feasible, but also all underlying assumptions for the hypersurface calculations can be checked in return. In addition, if both predicted and measured molecular properties coincide, the hypersurface areas between the local minima may provide information on isomers within a chemically reasonable energy range and the rearrangements involved. Short-lived intermediates which are by definition reactive have the special advantage, that - because of the prevailing energy transfer from the wall for low-pressure gas phase reactions and from the solvent in dilute solutions - no collisions have to be dealt with explicitly within this approximation. With their decomposition starting from a given structure, and in the case of rearrangements their constant composition, some clues are provided by hypersurface calculations as to which pathway(s) may be most probable. Despite all the caution recommended, this interplay of theoretical prediction and experimental finding thus can also contribute to improve the 'rather shallow understanding' (8) of the reactivity of medium-sized molecules.

In the following, six examples of hypersurface calculations and their correspondence to experimental results, will illustrate the benefit of this approach for the chemist dealing with short-lived intermediates (3,4).

Representative Examples of Semiempirical Hypersurface Calculations for Medium-Sized Molecules.

Six examples are selected to illustrate various facets of semiempirically calculated hypersurfaces for medium-sized molecules and their correlation with experiment. They comprise:

Example I: The gas phase structure of $H_3CS-C{\equiv}C-SCH_3$ and its ionization pattern.

Example II: The relative stabilities of C_3H_4S isomers and the thermal generation of $H_2C{=}CH-HC{=}S$.

Example III: The $N_2H_4{}^{\bullet\oplus}$ energy and coupling constant hypersurfaces and their application to approximate the structure of trialkylsilyl-substituted hydrazine radical cations in solution.

Example IV: The thermal and oxidative isomerization of tetraalkyl-substituted tetrahedrane clusters $(CR)_4$ to cyclobutadiene derivatives.

Example V: Existence and nonexistence of 1,1-substituted disulfide isomers $X_2S{=}S$ with X=F, H and Cl.

Example VI: Some aspects of the thermal decomposition of methyl and vinyl azides, H_3C-N_3 and $H_2C{=}CH-N_3$.

An outlook on what has been learned, where flaws are observed and what is needed from theoreticians to further improve the arsenal of methods available to the experimentalist to track short-lived intermediates and to gain insight in their reactions will conclude this survey based on own incomplete experience.

Example I: The Gas Phase Structure and the Ionization Pattern of Di(thiomethyl) Acetylene $H_3CS-C{\equiv}C-SCH_3$. Unstable substituted acetylenes can be generated in the gas phase by thermal decomposition of the corresponding cyclobutene-1,2-diones (14) splitting off 2 carbon monoxides as thermodynamically advantageous leaving groups (4):

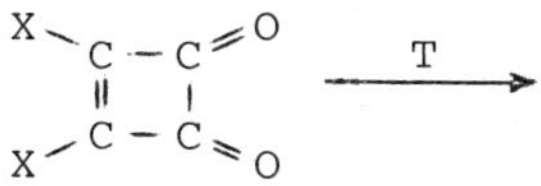

X-C≡C-X + 2 CO (1)

(X = Cl, SCH_3, $SeCH_3$)

The acetylenes generated - including the explosive dichloro derivative, which can be handled with minimum hazard under the decomposition conditions of 0.01 mbar helium atmosphere (4,14) - are identified by their photoelectron spectra, which in addition allow to optimize the reaction by following the vanishing ionization patterns of the starting material (4,14). In case of e.g. the di(thiomethyl) derivative, which exhibits 3 bands of double intensity in the low energy region, the correlation with hypersurface calculations to approximate its gas phase structure i.e. the dihedral angle between the 2 CS bonds, can be demonstrated in an exemplary way (Figure 1).

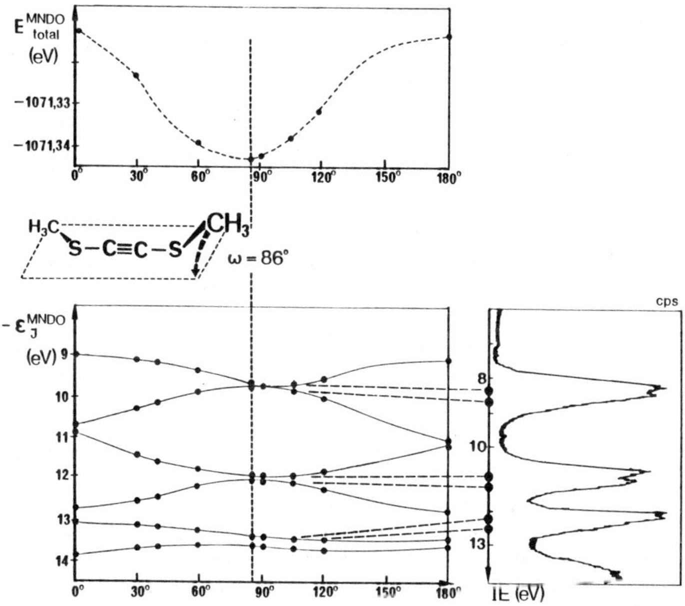

Figure 1. MNDO total energies and sequences of the 6 lowest MNDO eigenvalues for $H_3CS-C{\equiv}C-SCH_3$ depending on its dihedral angle ω(SC···CS) and correlation for ω = 86° with the PE spectroscopic ionization pattern.

Comparing the PE spectroscopic ionization pattern with the results of MNDO hypersurface calculations, at a first glance, satisfactory agreement is observed. The MNDO eigenvalue sequence obtained for the total energy minimum predicted at ω = 86° correlates linearly with the vertical ionization energies; the standard error for the Koopmans regression, $IE_n^V = 1.02(-\varepsilon_J^{MNDO}) - 1.51$, amounting to only SE = 0.18 eV. Moreover, the gas phase structure calculated reproduces the dihedral angle, $\omega^{exp.} = 86^o \pm 5^o$, determined in an independent electron diffraction study (15). As an additional advantage the time needed for the investigation might be pointed out: both the PE spectroscopic optimization of the thermal decomposition (1) and the MNDO hypersurface calculation were carried out each within a single day.

Some more general remarks, however, remain to be added concerning the accuracy of semiempirical calculations and the internal dynamics of the molecule investigated. A closer look at the energy scale of Figure 1 reveals that the minimum for the structure with almost perpendicular CS bonds is a rather shallow one - partly due to the assumed constant geometry for the H_3CS subunits. The rotational barrier calculated, $\Delta E_{total}^{MNDO} \sim 0.04$ eV $\simeq$ 4 kJ/mole, suggests that already at room temperature a propeller-like motion is activated. In a classical paper (16), Honegger and Heilbronner demonstrated via a numerical experiment including well over 100 rotational eigenfunctions and assuming free internal rotation that PE spectra of rotor molecules are correctly analyzed and can be simulated via the conformer distribution density.

Example II: The Relative Stabilities of C_4H_3S Isomers and the Thermal Generation and Subsequent Trapping of Thioacroleine. Hypersurface calculations are not only most valuable in rationalizing experimental results (cf. Example I), but often are quite helpful in the planning stage of investigations. Aiming, for example, at the preparation of unstable thiocarbonyl compounds (17-19), thioacroleine (18) is an attractive candidate. To elucidate chances of its PE spectroscopic detectability and thus to avoid unpromising experimental effort, all valence isomers of the ensemble C_4H_3S exhibiting the usual coordination numbers, $n_C \leqslant 4$ and $n_S \leqslant 2$, and connectivity patterns have been sorted out:

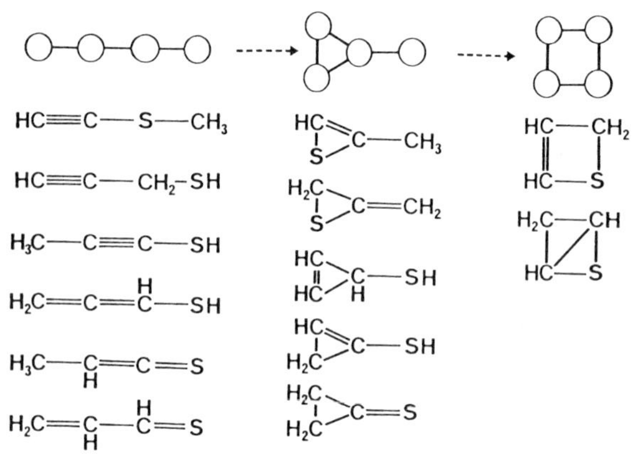

(2)

For the molecules (2) MNDO heats of formation are calculated by starting from known structural data and optimizing their respective geometries via the Davidson-Fletcher-Powell program subroutine (Figure 2).

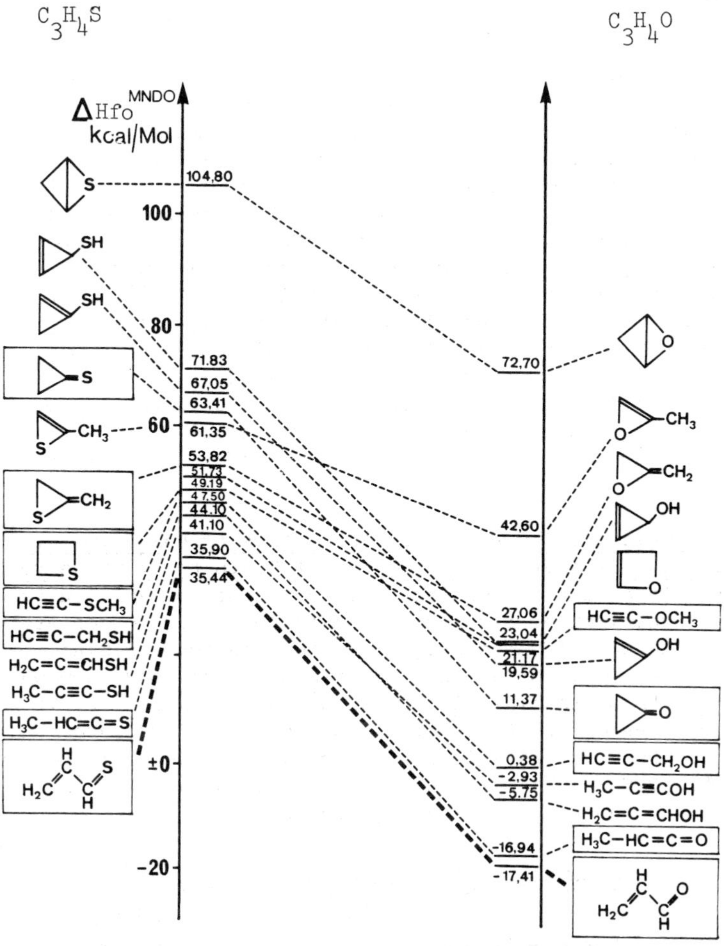

Figure 2. MNDO heats of formation for 'normal valence' isomers within the ensembles C_4H_3S and C_4H_3O (▭: compounds for which in 1980 literature references (17) could be found).

The MNDO screening for local minima in the C_4H_3S and C_4H_3O hypersurfaces yields a wealth of information (17), from which the following is emphasized here:

i. In both ensembles the isomers $H_2C{=}CH{-}HC{=}X$ are predicted to be the most stable ones.

ii. For all oxygen derivatives, considerably lower heats of

formation are calculated, e.g. the difference between acroleine and thioacroleine amounting to over 50 kcal/mole.

iii. Nevertheless, 6 isomers C_4H_3S of predicted lower relative thermodynamical stability were found reported in a subsequent literature search (17) (Figure 2: ▭).

From this hypersurface survey, incomplete with respect to approximate barriers for kinetic aspects e.g. concerning the ready polymerization of thiocarbonyl derivatives (19g), we tentatively concluded that the PE spectroscopic detection of thioacroleine (18) should be feasible, that it would be less stable than its oxygen analogue, acroleine, and that it could be identified by its PE spectroscopic ionization pattern because of the considerable differences in MNDO eigenvalue sequences to all 'nearby' C_4H_3S isomers (Figure 2).

The rest was routine: choosing propene as a thermodynamically favorable leaving group (17, 19, 20), diallylsulfide was synthesized despite its odor and pyrolyzed with PE spectroscopic optimization (4) of the conditions:

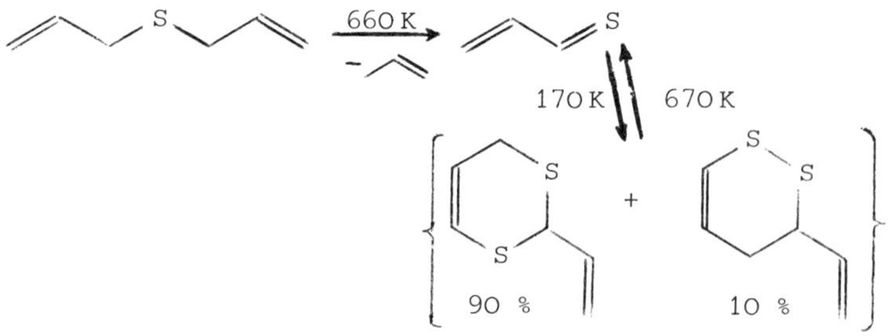

(2)

The violet thioacroleine readily dimerizes and can be isolated as a white crystalline mixture of 2-vinyl-1,3-dithiine(ΔHf^{MNDO} = 111 kJ/mole) and 3,4-dehydro-3-vinyl-1,2-dithiine (ΔHf^{MNDO} = 126 kJ/mole) in the NMR spectroscopically determined ratio 90:10 (17) - again agreeing well with the calculated ΔHf^{MNDO} difference. No indication is found for the third possible isomer, 3,4-dihydro-2-thioformyl-2H-thiopyrane, for which a much higher ΔHf^{MNDO} value is predicted. On evaporation, only the more stable isomer can be detected in the gas phase by its ionization pattern which correlates satisfactorily (17) with MNDO eigenvalues via Koopmans' theorem, $IE_n^V = -\varepsilon_J^{MNDO}$. Finally, thermal retrodiene splitting at 670 K yields pure thioacroleine, identified either by molecular state comparison with its oxygen analogue, acroleine, or based on the MNDO eigenvalue sequence, already calculated during the C_3H_4S isomer screening (Figure 2).

Example III: The Hydrazine Radical Cation Total Energy and Coupling Constants Hypersurfaces and Their Application to Approximate the Structure of Organosilicon Derivatives in Solution. For large molecules such as tetrakis (trimethylsilyl)hydrazine, $((H_3C)_3Si)_2N-N(Si(CH_3)_3)_2 = C_{12}H_{36}N_2Si_4$, or its radical cation

(21-23), containing 109 valence electrons, semiempirical calculations are still feasible and can reasonably reproduce some essential features even of peculiar compounds, e.g. of the azo derivative $(H_3C)_3Si-N=N-Si(CH_3)_3$ exhibiting a puzzling N=N bond length of only 117pm (21). In many other respects such as Koopmans' correlation with ionization energies, however, they are meaningless due to the growing gap between the numerical accuracy achievable by semiempirical procedures and the compression of many molecular orbitals into small ranges (5). If in addition, the computing effort is considered, the necessity to rather switch over to model calculations of the bare-laid backbone of the molecular skeleton becomes obvious.

The six-atom hydrazine radical cation, $H_2N\overset{\cdot\oplus}{-}NH_2$, possesses a total of 12 degrees of freedom. Contraction to 3 angular coordinates - the HNH bond angle α, the angle β for symmetrically bending an NH_2 group out of the plane defined by the NN bond, and the dihedral angle ω between the 2 amino groups, and assuming moreover fixed NH bond distances and an axis of rotation relative to which symmetrical intramolecular movements are defined - allows one to calculate a complete set of 2 hypersurfaces, each for a pair of coordinates (23,24). The INDO open shell procedure employed in this case also contained a Davidson-Fletcher-Powell subroutine, and thus each of the total 245 points calculated - varying α between 95° and 125° in 7 steps, β between 0° and 30° in 5 steps and ω between 0° and 180° in 7 steps - is fully geometry-optimized (CPU time: 2,5 hours on a UNIVAC 1108 (24)). Figure 3 presents one of the INDO total energy surfaces i.e. a symmetrically supplemented one for ω stretching from 0° to 360° with β as the more susceptible parameter, and in addition all 4 'hypersurface maps' for the angular dependence of the ESR hyperfine coupling constants, a_H and a_N.

The results of the careful hypersurface calculations were surprising: hydrazine with its dihedral angle ω=90° and an NN bond distance of 145 pm, on loss of one out of its 14 valence electrons, should form a completely planar (D_{2h}) radical cation with the NN bond length shrinking by 17 pm (!) to 128 pm (!). Luckily enough, we dared to publish this 'hard-to-believe' result (23), which a few months later has been completely confirmed by S.F. Nelson and collaborators (25), who succeeded in isolating crystals of the tetraalkyl hydrazine radical cation (3) and obtaining its X-ray structure, which exhibits an NN bond distance of 127 pm, i. e. close to the hypersurface prediction!

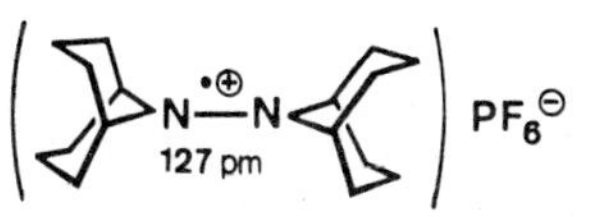

(3)

In the meantime, numerous hydrazine derivatives including those of boron (23,24), silicon (22 - 24,26) and phosphorus (23,24) had been oxidized using the powerful, selective and oxygen-free

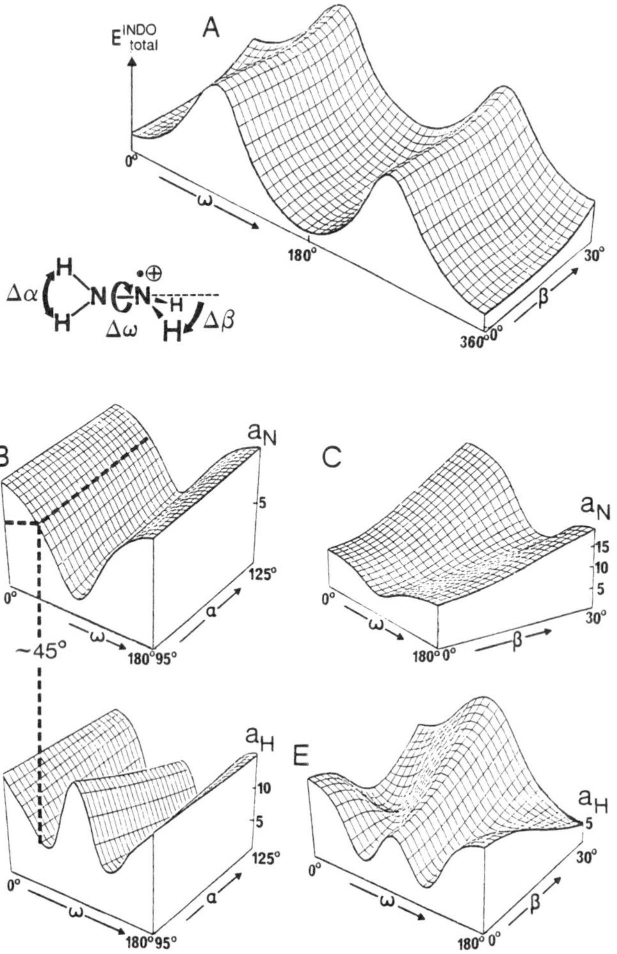

Figure 3. INDO open shell hypersurface calculations for the hydrazine radical cation (23,24) contracted to 3 angular coordinates of freedom (cf. text): (A) INDO total energies vs. the β/ω coordinate pair, and 'hypersurface maps' for the dependence of the ESR hyperfine coupling constants, a_N (B and C) and a_H (D and E) on the dihedral angle ω and the HNH bond angle α (B and D) or the out-of-plane bending angle β (C and E).

system $AlCl_3/H_2CCl_2$, based on straightforward predictions either by calculated lowest eigenvalues or by PE spectroscopically determined first ionization potentials (26). The assignment of the partly rather complicated ESR hyperfine splitting patterns comprising sometimes several thousand lines, however, was difficult to achieve. Even assuming considerable planarization did help only in simple cases, because there was hardly any information available on how and to what extent the prominent coupling constants would be affected by varying the individual angles. Again, the INDO open shell calculations for the model compound $N_2H_4^{\cdot\oplus}$ proved to be valuable (Figure 3): Looking at the hypersurface maps for the angular dependence of the coupling constants a_H and a_N, a change in the HNH bond angle α unexpectedly exerts no effect (diagrams A and C), varying β only a small one. It is the dihedral angle ω, which affects a_N and a_H most strongly and, surprisingly, in different ways: one minimum for a_N at $\omega \sim 90^o$, where a_H assumes a maximum value, and 2 minima for a_H at $\omega \sim 60$ and $\omega \sim 120^o$. To cut a longer story short: Comparing e.g. the coupling constants for planar bicyclic radical cation and for a open-chain tetraalkyl derivative (23,24):

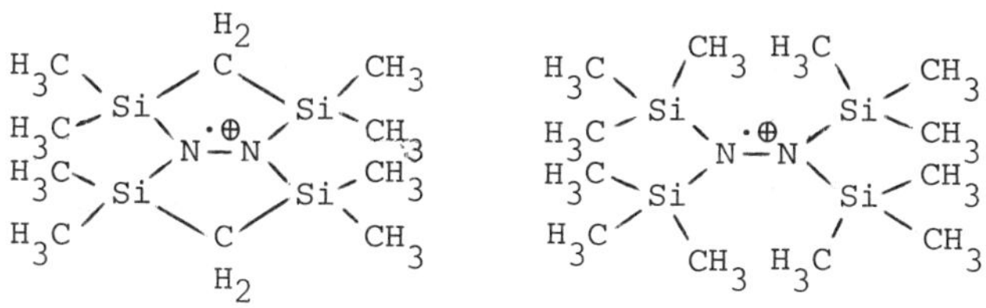

(4)

$I^{\cdot\oplus}$ $a_N = 0.868$ mT, $a_H = 0.50$ mT

$II^{\cdot\oplus}$ $a_N = 0.752$ mT, $a_H = 0.16$ mT

assuming proportionality of the spin density at the adjacent center, $a_H \sim a_{^{29}Si}$, and applying the calculated threedimensional hyperfine coupling constant tables (Figure 3: dotted line), both the a_N and a_H values suggest a dihedral angle $\omega \sim 45^o$ for $II^{\cdot\oplus}$. The reasoning implies that the open-chain silyl hydrazine radical cation should also be planar in its total energy minimum (Figure 3:A), but due to steric interference of the bulky trimethylsilyl substituents, the planarization stops about half-way.

Summarizing, this example provides several take-home lessons: complete sets of hypersurface calculations for main-frame models of compounds can be quite helpful in close correlation to experimental data. Obviously, both the radical cation ground state structure and the angular dependence of the coupling constants are correctly predicted. In return, by introducing experimental data into the established correlations, the structure of radical cations in solution may be cautiously approximated. Altogether, this example teaches another lesson on how drastic those structural changes may be, which accompany even one-electron redox reactions.

Example IV: The Thermal and Oxidative Isomerization of Tetraalkyl Substituted Tetrahedrane Clusters to Cyclobutadiene Derivatives. Other remarkable structural changes during redox reactions, i.e. charge redistributions enforced by the respective energy differences, are observed for cluster compounds (1). On addition of 2 electrons each to electron-deficient clusters of boron or the transition metals, usually the closo-spheres open to form nido- and, subsequently, arachno-baskets. On the other hand, electron withdrawal transforms e.g. the electron-rich S_4N_4 cage into the planar $S_4N_4^{\oplus\oplus}$ ring - predicted by hypersurface calculations and verified experimentally later on (27). Considering electron-saturated clusters such as the 'platonic' C_4 tetrahedron, what kind of structural changes will occur on oxidation?

Tetrahedrane $(CR)_4$ with bulky tert. butyl substituents R has been synthesized by G. Maier and coworkers in 1978 (28). In their already classic investigations, an interesting valence isomerization has been observed: on melting, the C_4 cluster rearranges to the thermodynamically more stable cyclobutadiene derivative, which on photochemical energy uptake forms again the tetrahedrane (28):

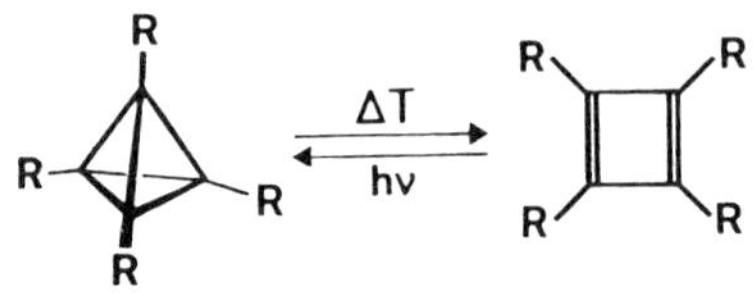

(5)

Both valence isomers easily give off an electron: the first vertical ionization energies amount to only 7.5 eV and 6.35 eV (29), and therefore, both compounds can be oxidized in solution using the selective one-electron oxidizing system $AlCl_3/H_2CCl_2$ (3). The result, however, is puzzling: in both cases, the tetrakis-(tert. butyl) cyclobutadiene radical cation is identified by its ESR spectrum (30).

In order to further substantiate the spectroscopic result and to gain more insight into both the thermal rearrangement of the neutral tetrahedrane and the structural change accompanying the one-electron oxidation to its radical cation, MNDO closed and open shell hypersurface calculations have been performed (30). The tetramethyl derivatives (5:$R=CH_3$) are chosen as model system to lower computational costs and, above all, to lay bare the electronic effects, which might be obscured by the steric requirement of the bulky tert. butyl groups (28,31). To further facilitate the calculations, a single isomerization coordinate α, along which the tetrahedron is flattened to a rectangle, has been chosen:

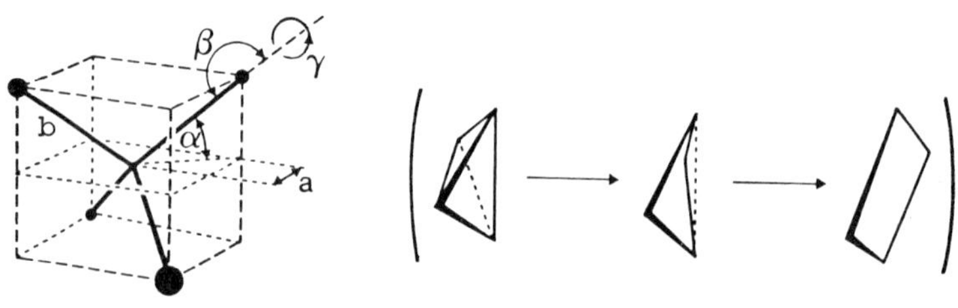

(6)

This choice of α defines an isomerization, in which D_2 symmetry is conserved along the rearrangement path and thus removes the restriction within $D_{2d} \rightarrow C_s$ or $D_{2d} \rightarrow C_2$ approaches, which would lead to symmetry-forbidden orbital crossings. Furthermore, the D_2 deformation corresponds closely (9) to the E-type normal vibrations in the ground state $X(^2E)$ of a tetrahedral radical cation which contribute to the PE spectroscopically observed Jahn Teller splitting (29). For the calculation itself, the CH_3 groups are assumed to remain constant, the parameters a (defining the flattening), b (CC bond distance) and the angles β and γ (6) are internally optimized, and the isomerization coordinate α is varied between 35° (regular tetrahedron) and 0° in 5° steps (Figure 4).

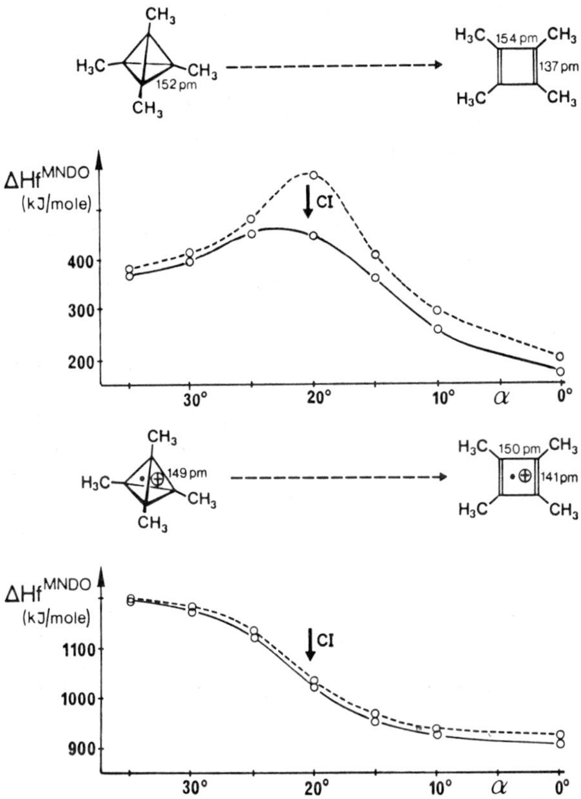

Figure 4. One-dimensional MNDO isomerization pathways for tetramethyl tetrahedrane and its radical cation. For the definition of the coordinate α cf. (6); CI denotes calculations including limited configuration interaction (see text).

The only one-dimensional MNDO cuts through the 3(20) - 6 = 54 dimensional hyperspace for $(C(CH_3))_4$ and its radical cation, nevertheless reproduce all essential experimental findings. They predict that the ring should be thermodynamically more stable than the cluster; the barrier is calculated with inclusion of a 10 x 10 configuration interaction to be approximately 80 kJ/mole compared to 107 kJ/mole determined experimentally (28), and to 64 kJ/mole calculated for the tert. butyl derivatives (31). Both the thermal cluster opening and the photochemical cluster formation can be rationalized straightforwardly. As concerns the oxidized cluster $(C(CH_3))_4^{\cdot\oplus}$, no barrier is calculated for the isomerization to the cyclobutadiene radical cation (Figure 4), in which according to MNDO the positive charge is distributed more evenly. The ESR observation, puzzling at first glance, that one-electron oxidation of both the tetra (tert. butyl) substituted tetrahedrane and cyclobutadiene yield the radical cation of the latter, thus finds a simple explanation.

It has to be pointed out, however, that additional valence isomers $(CR)_4$ such as cyclopropenyl carbene, are found in local energy minima, if other cuts through the 54 dimensional hyperspace are calculated (30) - in agreement with compounds isolated recently e.g. from the reactions of both isomers, tetra (tert. butyl) tetrahedrane and cyclobutadiene, with tetracyanoethylene (32).

Example V: The Rearrangement FSSF → $F_2S{=}S$ and the Predicted Nonexistence of $Cl_2S{=}S$ and $H_2S{=}S$. So far, conformers (Example I), isomers (Example II) and structural changes on one-electron oxidation (Example III and IV) have been dealt with. Another step in the direction of increasing complexity are those rearrangements, in which an atom is moved from one molecular site to another one, exemplified here by the isomerization of simple disulfides (33, 34):

(7)

(C_2; X = H,F,Cl) (C_s ; X = F)

For both the hydrogen and chlorine compounds, until now only the C_2 species are reported in the literature (33); whereas with fluorine substituents both isomers, difluorodisulfane as well as thiothionyl difluoride have been isolated and their structures determined by microwave spectroscopy (34). Surprisingly, the $F_2S{=}S$ molecule, which exhibits one of the shortest S=S bonds of onyl 189pm length, is thermodynamically more stable (34), and - even more - for both isomers, FSSF and $F_2S{=}S$, almost identical ionization patterns are observed (35). In order to rationalize these experimental results, and especially to find out,whether there are chances for the spectroscopist to also detect the C_s isomers, $H_2S{=}S$ and $Cl_2S{=}S$, hypersurface calculations were performed.

To reduce the 3(4)-6=6 dimensional problem to manageable 2

coordinates, atoms X have been moved stepwise by a program subroutine along a fixed SSX fragment, thereby crossing 12 perpendicular sectional planes of each 144 grid points. For the resulting 1728 spatial screening points CNDO total energies have been calculated, and additional geometry optimizations carried out around the detected minima. The results of this model approximation are displayed in Figure 5.

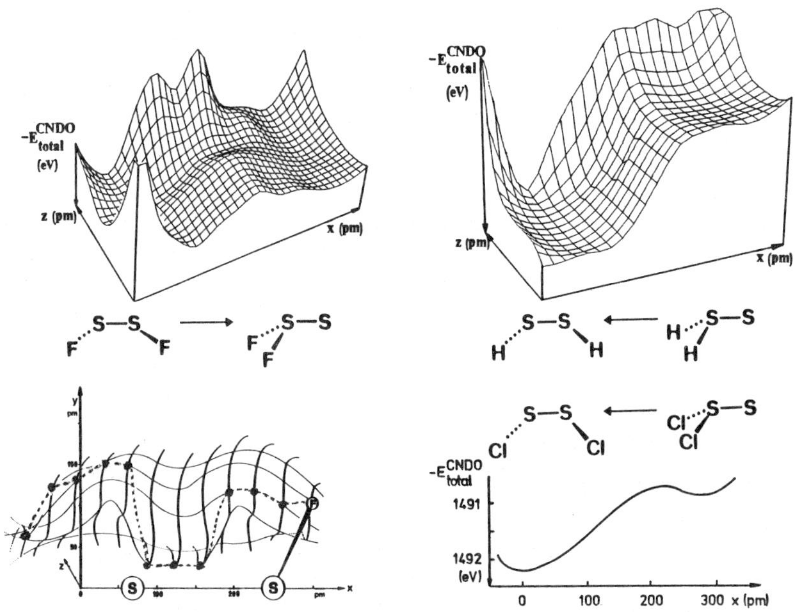

Figure 5. CNDO total energy hypersurface calculations for XSSX ⇄ X_2SS systems. Counterclockwise are displayed: the E^{CNDO}_{total} hypersurface for X=F and the approximate movement (-•-) of an F atom along a fixed SSF fragment on the bottom of the energy valley, the E^{CNDO}_{total} hypersurface for X=H, and the cut along the 'reaction coordinate' of lowest CNDO total energies for X=Cl.

The CNDO total energy surface for the F_2S_2 system reproduces in a qualitative way the relative thermodynamical stabilities observed experimentally i.e. it favors F_2S=S by $\Delta E^{CNDO}_{total} \sim 2$ kcal/mole with an upper barrier limit - obtained neglecting configuration interaction - of about 30 kcal/mole. Following the modelled movement of an F atom along the fixed SSF fragment through the 12 bisectional planes chosen (Figure 5: -•--), an intermediate is indicated: the F atom pathway winds around the S atom and crosses an almost planar three-membered ring before it reaches its final position. Whatever the reality of the isomerization process may be, the geometry optimized CNDO calculations around the energy minima for FSSF and F_2S=S correctly predict via Koopmans' theorem, $-\varepsilon^{CNDO} = IE^{V}_{n}$, that the structurally so different 4 atom valence isomers indeed should display almost identical ionization patterns as observed PE spectroscopically (35) - thereby supporting additional experimental results that the less stable FSSF has not been isomerized under the measurement conditions.

As concerns the H_2S_2 and Cl_2S_2 systems, the relatively large CNDO total energy differences between the isomers and the 'close to zero'-barriers calculated (Figure 5) strongly suggest the nonexistence of both $H_2S{=}S$ and $Cl_2S{=}S$ under usual circumstances.

Summarizing facts and speculations: despite the crudeness of the model chosen, the relative thermal stabilities of both isomers, FSSF and F_2SS are correctly reproduced by the CNDO hypersurface calculations, and also the surprising likeness of their ionization patterns (33-35). Whatever the real isomerization pathway may be, the chances for the experimentalist to ever isolate the molecules $H_2S{=}S$ and $Cl_2S{=}S$ under normal reaction conditions are predicted to be close to zero - another recommendation to avoid futile experimental efforts by precalculating hypersurfaces.

Example VI: Prediction and Detection of 2H-Azirine as Intermediate in the Thermal Decomposition of Vinyl Azide. Predictions concerning the hard-to-prove nonexistence of molecules are less convincing to the lay-colleague than those, which have stimulated successful experiments. Therefore, this last example has not only been selected because of its higher complexity, but rather due to the PE spectroscopic verification of the results anticipated by a preliminary hypersurface study (4,36).

Thermal decompositions of alkyl azides are advantageously studied in millimole quantities using a PE spectroscopically controlled flow system under low pressure (4), thereby reducing the hazards involved in handling these explosive compounds in bulk. Our investigations started with methyl azide, which splits off nitrogen unexpectedly only at temperatures above 500° C (37):

$$H_3C{-}N_3 \xrightarrow[-\ N_2]{850\,K} H_2C{=}NH \xrightarrow[-\ H_2]{} H{-}CN \qquad (8)$$

To rationalize both the high decomposition temperature and the immediate rearrangement to methane imine - no methyl nitrene intermediate can be detected under the measurement conditions -, hypersurface calculations have been performed. Assuming a dissoziation into singulet fragments, the $H_3CN{-}NN$ bond has been lengthened in 20 pm steps along the direction of N-N vibrational stretching frequency under geometry optimization of both the H_3CN and N_2 subunits: the upper-limit barrier obtained (neglecting configuration interaction) of $\Delta Hf^{MNDO} \sim 220$ kJ/mole is reasonably close to the experimental estimate of $\sim$ 170 kJ/mole (37). An MNDO hypersurface for the rearrangement of singulet methylnitrene to the observed methane imine showed a barrier of only $\sim$ 10 kJ/mole.

Based on this qualitatively satisfying agreement between PE

spectroscopic observation and MNDO approximation, we tackled the next problem envisaged, the thermal decompositon of the even more dangerous vinylazide (38), by precalculating an MNDO hypersurface for vinyl nitrene, $H_2C=CH-N$, assumed in close analogy to be an undetectable short-lived intermediate. Pin-point exploration for normal valence isomers (cf. Example II) within the C_2H_3N ensemble resulted in the following order of increasing MNDO total energies: starting from H_3C-CN as the most stable species via $H_2C=C=NH$, H_3C-NC and $HC\equiv C-NH_2$, the three-membered ring, 2H-azirine, is the last molecule predicted to exist within a chemically feasible energy range (36). Finally, reducing the 3(6)-6=12 dimensional problem to the 2 selected angular coordinates, ∢ CCH and ∢ CCN, an MNDO hypersurface could be calculated which comprised at least 4 of the essential isomers on the same energy plane (Figure 6).

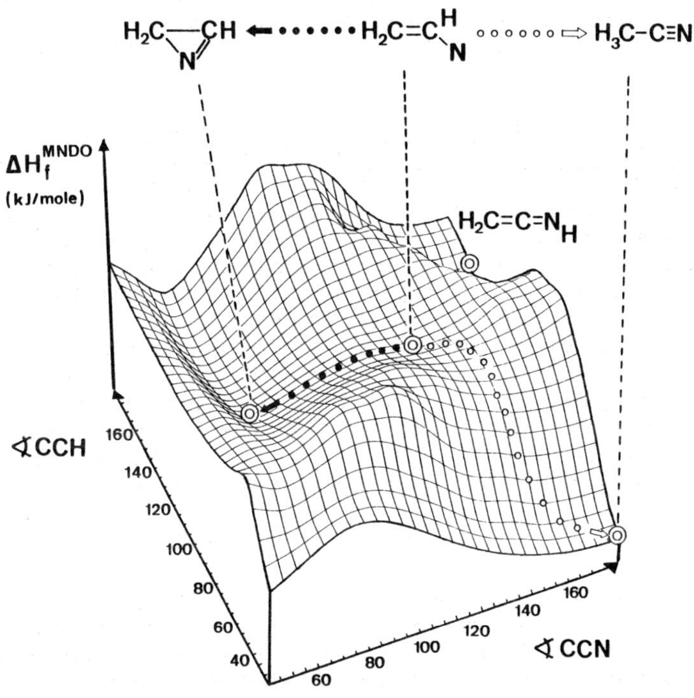

Figure 6. MNDO heat of formation hypersurface for singlet vinyl nitrene rearrangements to 2H-azirine, ketene imine and the thermodynamically most favorable isomer, acetonitrile.

Inspection of the MNDO hypersurface shows the assumed vinyl nitrene to lie in a very shallow minimum with the activation barrier for ring closure to 2H-azirine (Figure 1: ••• ➡) calculated to be only a few kJ/mole. Whereas the separating hypersurface "ridge" to ketene imine of $\Delta\Delta$ Hf^{MNDO} ∿ 70 kJ/mole seems insurmountable, the rearrangement to acetonitrile (Figure 1: ooo ⇒) can proceed along a more favorable "mountain pass" of $\Delta\Delta Hf^{MNDO}$ $\sim$ 40 kJ/mole (39).

The hypersurface prediction (Figure 6) has been verified by the following experiments: Vinyl azide, which is described to detonate at only 70° C (38), has been synthesized in ten millimole quantity following the literature procedure (38) and vaporized without decompostion from a 195 K cooling trap at 10^{-4} mbar. Its PE spectroscopically controlled pyrolysis over quartz wool starts at an oven temperature of 620 K. Freezing out all condensable decomposition products in a U-tube cooled to 77 K, followed by subsequent fractioned vaporization allows to record the PE spectrum of pure 2H-azirine, assigned by Koopmans' correlation, $IE_n^v = -\varepsilon_J^{MNDO}$, with the eigenvalue sequence from the geometry-optimized MNDO calculations (4,37). Substituting the U-trap by a second oven heated to temperatures above 650 K, the continuously registered PE spectra show increasing intensities of the well-known ionization pattern of acetonitrile (37) - predicted as the most stable isomer of the C_2H_3N ensemble in the preceeding hypersurface study (Figure 6).

This last example again demonstrates the usefulness of correlating hypersurface calculations as closely as possible to measurement data: although methyl and vinyl nitrenes, which by chemical intuition seem to be reasonable intermediates, do not have to occur necessarily along the reaction pathway (36,39), the predictions stimulated an experiment, which otherwise - due to the hazardous starting material (38) - hardly would have been carried out.

Concluding Remarks

The Examples I to VI have been presented to demonstrate the benefit of close correlation between hypersurface calculation and measurement to the experimentalist. In the order of increasing complexity have been discussed: conformer (I) and isomer (II) stabilities, structural changes following oxidation (III,IV), valence isomerization (IX), one-step rearrangements (V) and a three-step thermal decomposition (VI). The hypersurface cuts were calculated using closed- (I,II,IV-VI) and open-shell (III, IV) versions of semiemperical MNDO (I,II,IV,VI), INDO (III) and CNDO (V) procedures, partly including configuration interaction (IV). The (3n-6) dimensional hyperspaces for ensembles (II,III, V,VI) or subunits thereof (I,IV) with n centers have been reduced - predominantly by transfer of Cartesian coordinates into angular ones, and by making use of symmetry - to hypersurfaces of dimensions 1 (I:n=4, IV:n=4), 2 (V:n=4, VI:n=6) and 3 (III:n =6), partly optimizing other structural features via the Davidson-Fletcher-Powell program subroutine. Satisfactory agreement with the following experimental data has been quoted: structural parameters (I, III, V), ionization patterns (I,II,VI), ESR coupling constants (III,IV), energy barriers (IV,VI), and - last but not least - compounds isolated (II,IV,V,VI). The usefulness of predicted 'molecular fingerprints' to detect and characterize especially short-lived molecules has been emphasized in Examples I, II and VI.

Numerous other examples have been reported in the literature by other groups - predominantly covering carbon skeleton rearrangements (40),and, using highly correlated wavefunctions, small molecules (7). As concerns semiempirical hypersurface calculations for medium-sized molecules containing third-row hetero elements, own experience in addition to Examples I, II and V can be summarized as follows: inclusion of elements such as P(41,42) and S (43-45) usually leads to satisfactory correlation with experimental data; in the special case of $H_3CSe\text{-}SeCH_3$ a properly parametrized EHMO version could be applied to achieve approximate results (46).

For a complete picture, however, some critical remarks have to be added. For medium-sized molecules, the obviously successful combination of theoretical prediction and experimental verification rather should be considered as an individually tailored interplay, which is based on numerous assumptions i.e. so-called 'chemical intuition'. To begin with, any hypersurface design confined to a few selected coordinates will only cover the aspects introduced thereby. Even several independent cuts through the (3n-6) dimensional hyperspace cannot unravel the complexity of a molecule in a specific molecular state. And although semiempirical methods - especially MNDO (10)- are not only fast but to quite an extent also reliable, their numerical accuracy should not be overstressed. Thus semiempirical hypersurfaces might better be considered as supplying trends in essential features, which can be compared to and correlated with measurement data to add the numerical accuracy, and, in return, to test and to substantiate all underlying assumptions.

In general, energy minima on a given hypersurface can be localized straightforwardly using semiempirical procedures; and the molecular properties predicted after appropriate optimization such as structural features or ionization patterns usually turn out to be close and reliable approximations of experimental facts. For example, structural changes accompanying redox reactions are satisfactorily reproduced by independent calculations of the ground states of both the neutral molecule and the radical ion generated, i.e. by the local minima on 2 different hypersurfaces. On the other hand, regions outside the local minima on a given hypersurface suffer from computational as well as from conceptual uncertainties. This applies especially to the transition 'antistates' i.e. saddle points of maximum energy with respect to certain coordinate changes through which a transforming molecule will pass most rapidly on its way from one local minimum to another one. Even if - as discussed repeatedly referring to (6) - a specific saddle-point is unambiguously identified and the approaches from both adjacent minima fortuitously meet there on forward and backward computation, still only those out of the 3n-6 degrees of freedom are accounted for, which have been specified in the hypersurface design. In order to obtain closer information on the actual pathway of reactions (8), improvement of semiemperical approaches by more specifically including certain features of the molecular dynamics involved would be a worthwile goal for theoreticians; much appreciated by the experimentalists.

Nevertheless, gratefully using what is already available from the quantum chemical cabinet, and under lucky circumstances, the preparative chemist can obtain considerable help in planning some of his daily work more properly - as outlined e.g. for the thermal decomposition of the hazardous vinyl azide in Example VI. And, as the last concluding remark, it may be appropriate to point out rather generally: a bit more familiarity with certain facets of the molecular state picture such as the intricate relationship between charge distribution and energy and the consequences thereof (5), by all means will provide an advantageous rationale to design experiments, to interprete results and to improve our insight into the sophisticated behaviour of molecules — or as the physicist and philosopher G. C. Lichtenberg expressed it almost 200 years ago: 'In nature we see not words but only the first letters of words, and when we wish to read, we find that the new so-called words are once again mere first letters of other words' (47).

Acknowledgments

Work of the Frankfurt Photoelectron Spectroscopic Group cited above would not have been possible without the effort of dedicated coworkers, especially Doz. Dr. Pavel Rosmus - whose discussion remarks on hypersurfaces are gratefully appreciated -, Dr. Bahman Solouki, Dr. Andrzej Semkow, Dr. Udo Stein, Dr. Shamsher Mohmand, Dr. Takakuni Haribayashi, Doz. Dr. Wolfgang Kaim and Dr. Sitki Aygen. Generous support of our research by the State of Hesse, the Deutsche Forschungsgemeinschaft, the Stiftung Volkswagenwerk, the Fonds der Chemischen Industrie, the Max-Buchner- and Hermann-Schlosser-Stiftung is gratefully acknowledged. Scientific contact with numerous colleagues have always had a stimulating effect, for cooperation cited here, we wish to thank expecially Prof. G. Fritz (Karlsruhe) and Prof. G. Maier (Giessen) as well as Prof. E. Block (Albany, N.Y.), Prof. J.W. Conolly (Kansas City), Prof. L. Horner (Mainz), Prof. H. Noeth (München), Prof. W. Ried (Frankfurt), Prof. F. Seel (Saarbrücken), and - above all - Prof. M.J.S. Dewar (Austin, Texas), who not only supplied the MNDO version used but also transferred much of his enthusiasm concerning hypersurface calculations to our group.

Literature Cited

1. Presented in part at the 'Ralph W. Rudolph Memorial Symposium' at the ACS Meeting, Kansas City, on September 15th, 1982. For a summary of some of the numerous valuable contributions of the late R.W. Rudolph to Inorganic Chemistry cf. "Borane Clusters, Metal Clusters and Catalysis" in Parry, R.W.; Kodama, G., Ed.; "Boron Chemistry", Pergamon Press: Oxford and New York 1980; p. 11-22, and literature cited, expecially Wade, K. Adv. Inorg. Chem. Radiochem. 1976, 18, 1.
2. 10th Essay on Molecular Properties and Models. For the preceding 9th on radical ions cf. (3) and the 8th on the optimization of gas phase reactions using photoelectron spectroscopy cf. (4).
3. Bock, H.; Kaim, W. Acc. Chem. Res. 1982, 15, 9.
4. Bock, H.; Solouki, B. Angew. Chem. 1981, 93, 425; Angew. Chem. Int. Ed. 1981, 20, 427; with 142 literature quotations.
5. Bock, H. Angew. Chem. 1977, 89, 631; Angew. Chem. Int. Ed. 1977, 16, 613; and literature quoted.
6. Cf. e.g. Müller, K. Angew. Chem. 1980, 92, 1; Angew. Chem. Int. Ed. 1980, 19, 1; and literature quoted.
7. Cf. also Bernstein, R.B., Ed.; "Atom-Molecule Collision Theory"; Plenum Press New York and London 1978, with special reference to Schaefer, H.F. III; Kuntz, P.J.; "Interaction Potentials", and the literature quoted.
8. Quoted from an interview with G. Pimentel as NSF Deputy Director on forecast advances for 1980's, cf. Chem. Engng. News (Jan. 14) 1980, p.7.
9. For a more qualitative approach based on symmetry cf. Halevi, E.A. Angew. Chem. 1976, 88, 664; Angew. Chem. Int. Ed. 1976, 15, 593.
10. Dewar, M.J.S.; Thiel, W. J. Am. Chem Soc. 1977, 99, 4907. The usefulness of MNDO type calculations to approximate hypersurfaces is pointed out e.g. by Jug, K. Theor. Chim. Acta 1980, 54, 263.
11. Cf. e.g. Kutzelnigg, W. "Einführung in die Theoretische Chemie", Vol. 1: "Quantenmechanische Grundlagen", Verlag Chemie: Weinheim 1975.
12. Cf. e.g. the Annual Review on Theoretical Chemistry in Nachr. Chem. Techn. Lab., but also (10).
13. Cf. e.g. Davidson, R.A. J. Am. Chem. Soc. 1981, 103, 312, and literature quoted.
14. Bock, H.; Ried, W.; Stein, U. Chem. Ber. 1981, 114, 673.
15. Beagley, B.; Ulbrecht, V.; Katsumata, S.; Lloyd, D.R.; Connor, J.A.; Hudson, G.A. J. Chem. Soc. Faraday Trans. II 1977, 73, 1278.
16. Honegger, E.; Heilbronner, E. Chem. Phys. Lett. 1981, 81, 615.
17. Bock, H.; Mohmand, S.; Hirabayashi, T.; Semkow, A. Chem. Ber. 1982, 115, 1339; preliminary communication: J. Am. Chem. Soc. 1982, 104, 312.

18. Cf. also Georgiu, K.; Kroto, H.W. J. Mol. Spectrosc. 1980, 83, 94.
19. Cf. e.g. (a) Bock, H.; Hirabayashi, T.; Mohmand, S. Chem. Ber. 1982, 115, 492; see also (b) Solouki, B.; Rosmus, P.; Bock, H. J. Am. Chem. Soc. 1976, 98, 6054; (c) Bock, H.; Solouki, B.; Bert, G.; Rosmus, P. J. Am. Chem. Soc. 1977, 99, 1663; (d) Rosmus, P.; Solouki, B.; Bock, H. Chem. Phys. 1977, 22, 453; (e) Bock, H.; Solouki, B.; Mohmand, S.; Block, E.; Revelle, L.K. J. Chem. Soc. Chem. Comm. 1977, 287; (f) Bock, H.; Aygen, S.; Rosmus, P.; Solouki, B. Chem. Ber. 1980, 3187; (g) Mohmand, S.; Bock, H. Phosphorus and Sulfur 1982, ..., ...; or (h) Bock, H.; Mohmand, S.; Hirabayashi, T.; Maier, G.; Reissenauer, H.P. Chem. Ber. 1983, 116, 273.
20. The propene split-off route to generate intermediates is well-established, cf. Brown, R.F.C. "Pyrolytic Methods in Organic Chemistry"; Academic Press: New York 1980. It has been applied by our group to prepare e.g. silatoluene (Bock, H.; Bowling, R.A.; Solouki, B.; Barton, T.J.; Burns, G.T. J. Am. Chem. Soc. 1980, 102, 429), silabenzene (Solouki, B.; Rosmus, P.; Bock, H.; Maier, G. Angew. Chem. 1980, 92, 56; Angew. Chem. Int. Ed. 1980, 19, 51), as well as numerous thiocarbonyl derivatives including thioformaldehyde (17a) or 1,4-dithiobenzoquinone (17h).
21. Bock, H.; Wittel,K.;, Veith, M.; Wiberg, N. J. Am. Chem. Soc. 1976, 98, 109.
22. Bock, H.; Kaim, W.; Connolly, J.W. Angew. Chem. 1976, 88, 766; Angew. Chem. Int. Ed. 1976, 15, 700.
23. Bock, H.; Kaim, W.; Semkow, A.; Nöth, H. Angew. Chem. 1978, 90, 308; Angew. Chem. Int. Ed. 1978, 17, 286.
24. Bock, H.; Kaim, W.; Nöth, H.; Semkow,A. J. Am. Chem. Soc. 1980, 102, 4421, and literature quoted, especially the survey given on the fine contributions by Nelsen, S.F. and coworkers.
25. Nelsen, S.F.; Hollinsed, W.C.; Kessel, C.R.; Calabrese, J.C. J. Am. Chem. Soc. 1978, 100, 7876.
26. Cf. also the review on "Organosilicon Radical Cations" by Bock, H.; Kaim, W. Acc. Chem. Res. 1982, 15, 9.
27. Gleiter, R. J. Chem. Soc. A 1970, 3174; and Gillespie, R.J.; Shin, D.R.; Tyrer, J.D. J. Chem. Soc. Chem. Comm. 1977, 253.
28. Cf. e.g. Maier, G.; Pfriem, S.; Schäfer, U.; Matusch, R. Angew. Chem. 1978, 90, 552; Angew. Chem. Int. Ed. 1978, 17, 520; Maier, G. Tetrahedron Lett. 1978, 1837; Maier, G.; Pfriem, S.; Schäfer, U.; Malsch, K.D.; Matusch, R. Chem. Ber. 1981, 114, 3965; or Maier, G.; Pfriem, S.; Malsch, K.D.; Kalinowski, H.O.; Dehnicke, K. Chem. Ber. 1981, 114, 3988, as well as literature quoted.
29. Heilbronner, E.; Jones, T.B.; Krebs, A.; Maier, G.; Malsch, K.D.; Pocklington, J.; Schmelzer, A. J. Am. Chem. Soc. 1980, 102, 564.
30. Bock, H.; Roth, B.; Maier, G. Angew. Chem. 1980, 92, 213; Angew. Chem. Int. Ed. 1980, 19, 209, as well as literature quoted.

31. Cf. Schweig, A.; Thiel, W. J. Am. Chem. Soc. 1979, 101, 4742.
32. Maier, G.; Schneider, K.A.; Malsch, K.D.; Irmgartinger, H.; Lenz, A. Angew. Chem. 1982, 94, 446, Angew. Chem. Int. Ed. 1982, 21, 437.
33. Solouki, B.; Bock, H. Inorg. Chem. 1977, 16, 665, and literature quoted.
34. For the F_SS_2 isomers cf. Seel, F.; Adv. Inorg. Chem. Radiochem. 1974, 16, 297; and Kuczkowski, R.L. J. Am. Chem. Soc. 1964, 86, 3617; 1963, 85, 3407.
35. Wagner, G.; Bock, H.; Budenz, R.; Seel, F. Chem. Ber. 1973, 106, 1285.
36. Bock, H.; Aygen, S.; Dammel, R., submitted for publication. Cf. Ph. D. thesis Aygen, S.: University of Frankfurt 1982.
37. Bock, H.; Dammel, R. Horner, L. Chem. Ber. 1981, 114, 220, and literature quoted.
38. Cf. e.g. Wiley, R.H.; Moffat, J. J. org. Chem. 1957, 22, 995; and literature quoted.
39. Independent ab initio calculations by Lohr, L.N. (private communication) would favor a reaction pathway via H_3C-NC not considered in Figure 6. For a general discussion of vinyl azide decomposition cf. e.g. L'abbé, G. Angew. Chem. Int. Ed. 1975, 14, 775; and for 1,2-hydrogen shifts cf. e.g. Schaefer, H.F. III Acc. Chem. Res. 1979, 12, 288, and the literature quoted in each.
40. For a compilation up to 1980 cf. e.g. Ing. K. Theor. Chim. Acta 1980, 54, 263.
41. Ostoja Starzewski, K.A.; Bock, H. J. Am. Chem. Soc.1976, 98, 8486.
42. Bock, H.; Müller,H.; submitted for publication; cf. (4).
43. Bock,H.; Stein, U.; Semkow, A. Chem. Ber. 1980, 113, 3208.
44. Bock, H.; Schulz, W.; Stein, U. Chem. Ber. 1981, 114, 2632.
45. Bock, H.; Schulz, W.; Schmidt, M. Z. anorg. allg. Chem. 1981, 474, 199.
46. Tschmutowa, G.; Bock, H. Z. Naturforsch. 1976, 316, 1616.
47. Between 1789 and 1793, Georg Christoph Lichtenberg made the quoted entry in his 'Sudelheft I' (cf. 5).

RECEIVED March 30, 1983

10

Synthesis of Phosphorus–Nitrogen Polymers by Using Silicon–Nitrogen–Phosphorus Reagents

PATTY WISIAN-NEILSON, AROOP K. ROY, ZE-MIN XIE, and ROBERT H. NEILSON

Texas Christian University, Department of Chemistry, Fort Worth, TX 76129

Many small-molecule Si-N-P compounds of appropriate design can function as precursors to various oligomeric and polymeric P-N systems. For example, poly(dialkylphosphazenes), $(R_2PN)_n$, are readily obtained via the thermal condensation-polymerization of certain N-silylphosphinimines, $Me_3SiN{=}P(X)R_2$. In addition, novel 3-coordinate phosphorus-V compounds including the bis(imino)-phosphoranes, $RP({=}NSiMe_3)_2$, can potentially lead to new polymers of formula $(RPN)_n$, the phosphorus analogs of electrically conducting $(SN)_n$. The synthesis of polyphosphazenes and their precursors and the synthesis of potential precursors to $(RPN)_n$ are discussed.

The structural and chemical diversity of compounds containing the Si-N-P linkage is a result of both the variety of coordination numbers which phosphorus may assume and the reactivity of the Si-N bond. The combination of these features gives Si-N-P compounds very promising synthetic utility, particularly in the area of inorganic polymer synthesis.

Typically, the most common precursors to new Si-N-P systems are simple silylaminophosphines (eq 1). The difunctional character of these compounds, which is due to the nucleophilic site at phosphorus and a complementary electrophilic site at silicon, makes them very versatile reagents. They have been used in a new synthesis of alkyl and/or phenyl substituted phosphazenes $(R_2PN)_n$ (1) and have led to the preparation of promising precursors to potentially electrically conducting polymer systems of general formula $(RPN)_n$.

0097–6156/83/0232–0167 $06.00/0

Synthesis of Polyphosphazenes

Work in the polymer synthesis area has been enhanced by a convenient "one-pot" synthesis which was developed by J.C. Wilburn (2) (eq 1). Yields of the phosphines prepared in this manner are typically 60 to

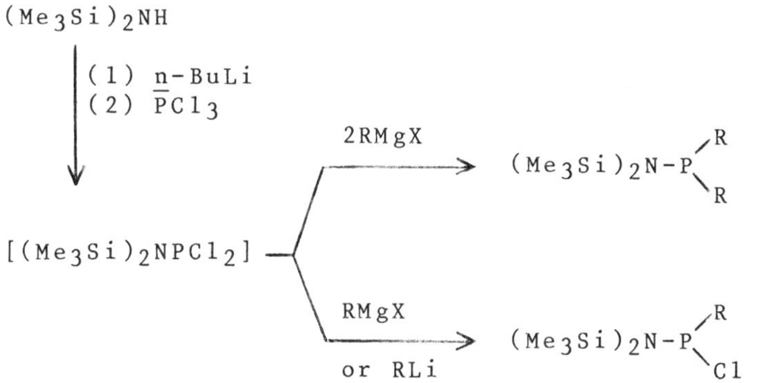

(1)

to 90% and quantities of 250 to 300 grams are easily obtained. The use of $PhPCl_2$ in place of PCl_3 allows for the introduction of phenyl/alkyl substitution at phosphorus.

Many simple oxidation reactions of tertiary alkyl or aryl phosphines are less straightforward in phosphines with silylamino substituents (3-8). Two such reactions have resulted in precursors to phosphazenes. The first of these is the reaction of bromine with disubstituted silylaminophosphines (7) to give Me_3SiBr and P-bromo-N-silylphosphinimines (eq 2).

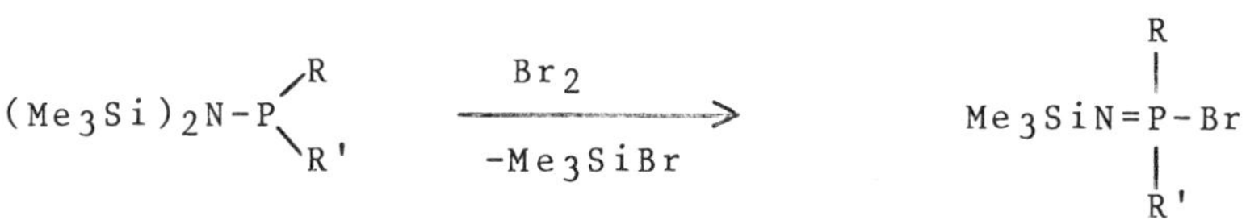

(2)

R,R' = Me, Et, i-Pr, Ph, OCH_2CF_3, CH_2Ph,

CH_2SiMe_3, $CH_2CH=CH_2$, etc.

The phosphinimines are thermally unstable and readily eliminate Me_3SiBr at temperatures of greater than 100°C to form only cyclic phosphazenes (eq 3). (7)

$$Me_3SiN{=}\overset{R}{\underset{R'}{P}}{-}Br \xrightarrow{\Delta} \left[\overset{R}{\underset{R}{P}}{=}N\right]_n + Me_3SiBr \qquad (3)$$

$$n = 3,4$$

More importantly, the P-bromophosphinimines react smoothly with CF_3CH_2OH in the presence of Et_3N to form P-(trifluoroethoxy)phosphinimines (eq 4) (7). On heating in either sealed glass ampoules or a stainless

$$Me_3SiN{=}\overset{R}{\underset{R'}{P}}{-}Br \xrightarrow[Et_3N]{CF_3CH_2OH} Me_3SiN{=}\overset{R}{\underset{R'}{P}}{-}OCH_2CF_3 \qquad (4)$$

steel bomb these phosphinimines readily eliminate $Me_3SiOCH_2CF_3$ (eq 5) and form exclusively polymeric phosphazenes. (1)

$$Me_3SiN{=}\overset{R}{\underset{R'}{P}}{-}OCH_2CF_3 \xrightarrow{\Delta} \left[\overset{R}{\underset{R'}{P}}{=}N\right]_n + Me_3SiOCH_2CF_3 \qquad (5)$$

R	Me	Et	Ph	Ph	Ph
R'	Me	Et	Me	Et	Ch_2Ph

The physical properties of these polymers vary considerably as the substituents at phosphorus are changed. Poly(dimethylphosphazene) $(Me_2PN)_n$ is a white film-forming polymer with a weight averaged molecular weight ($\bar{M}_w$) of 50,000 as determined by light scattering. It is soluble in CH_2Cl_2, $CHCl_3$, CH_3CH_2OH, and THF/H_2O, and has a melting point of 158°C and a glass transition temperature of -42°C (1). The diethyl analog $(Et_2PN)_n$, on the other hand, is quite perplexing since its virtual insolubility in all solvents precludes most characterization other than elemental analysis.

In contrast to the alkyl substituted polymers, the P-phenyl compounds, $[Ph(Me)PN]_n$ and $[Ph(Et)PN]_n$, are brittle, brown colored materials which are quite soluble in THF and are readily plasticized by trace amounts of solvent. Determination of molecular weight by gel permeation chromatography for $[Ph(Me)PN]_n$ indicates $\bar{M}_n$ to be 54,000. Its glass transition temperature was found to be 37°C.

Phosphazene copolymers have also been prepared by the thermolysis of equimolar mixtures of the phenyl/alkyl and dialkyl substituted P-trifluoroethoxyphosphinimines.

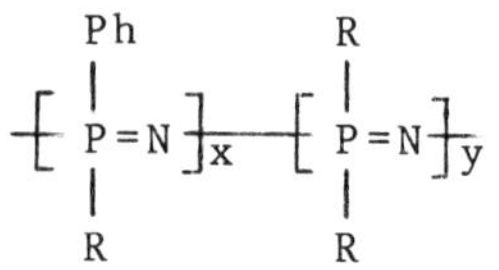

R = Me, Et

These materials are generally more elastomeric and rubber-like in appearance than either of the corresponding homopolymers. Such systems indicate that it should be possible to "custom design" polyphosphazenes by choosing precursors with desirable substituents.

In summary, this new and general method of synthesizing both cyclic and polymeric phosphazenes (eq 6)

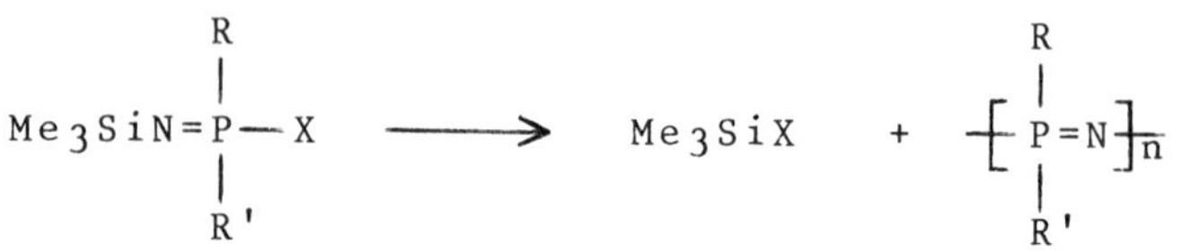

(6)

involves properly designed N-silylphosphinimines which contain a leaving group X and the desired substituents R and R' on phosphorus. Such compounds eliminate substituted silanes Me_3SiX and form cyclic or polymeric phosphazenes. When X is Br or, as reported earlier (9), F, only small ring compounds are formed, but when X is CF_3CH_2O, polymeric phosphazenes result. In contrast to the usual methods of preparing inorganic polymers via ring opening polymerization, this is an unusual example of a condensation-polymerization which gives an inorganic P-N polymer backbone. This method, therefore, has the advantage of allowing the initial construction of small molecule building blocks which incorporate desirable side groups. This is particularly useful for the preparation of alkyl and/or aryl substituted polyphosphazenes with direct carbon to phosphorus bonds, a system which is difficult to achieve by other methods.

CCl_4 Reactions of (Silylamino)phosphines

Other N-silylphosphinimines which are also suitably constructed phosphazene precursors, can be prepared by a second oxidation reaction of silylaminophosphines. When these phosphines react with carbon tetrachloride, two types of P-chlorophosphiminines may be formed by essentially two pathways (eq 7) (8).

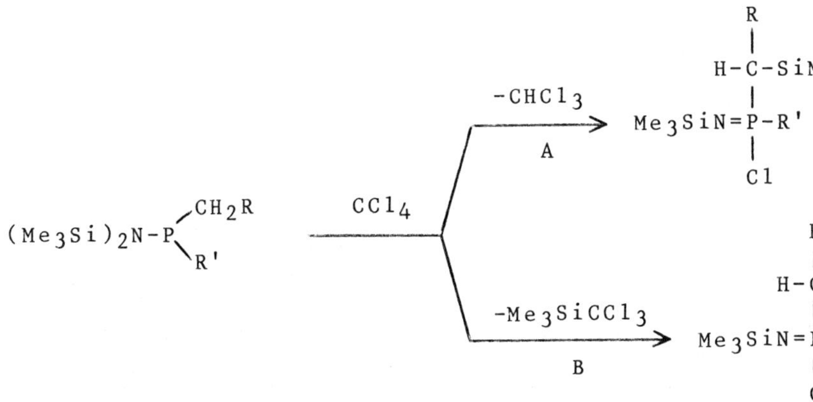

(7)

R = $SiMe_3$, Me, Ph, H
R' = CH_2SiMe_3, Me, Et, *i*-Pr, *t*-Bu, Ph, NMe_2, OMe, OCH_2CF_3

The elimination of $CHCl_3$ as shown in pathway A occurs in similar systems which contain a C-H moiety α to phosphorus. Presumably, initial formation of an ion pair intermediate $(R_3PCl^+)(CCl_3^-)$ (10) is followed by attack of the CCl_3^- anion at the hydrogen on the α carbon with elimination of $CHCl_3$ giving ylide products (eq 8) (11,12). However, in systems which

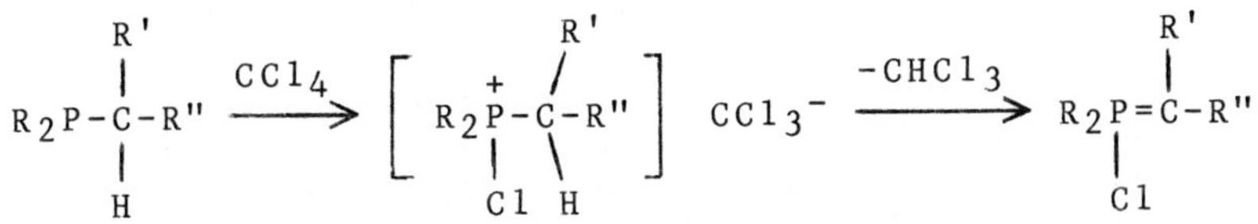

(8)

contain a silylamino substituent, the analogous ylides are presumably intermediates which rearrange via a [1,3]-silyl shift (eq 9) (4) to give the P-chlorophosphinimines shown in pathway A.

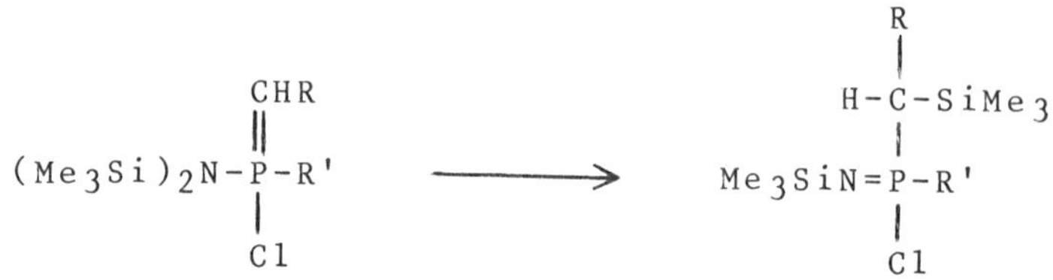

(9)

An alternate pathway is possible for systems containing silylamino substituents at phosphorus. This most likely involves attack of the CCl_3^- anion at the electrophilic silicon resulting in elimination of Me_3SiCCl_3 as shown in pathway B. In the systems investigated thus far, the reaction pathway preference appears to be influenced by: (1) solvent polarity, and (2) steric and electronic effects of the substituents at phosphorus (8).

Low-coordinate Phosphorus Systems

Regardless of which N-silylphosphinimine products are formed they all are potential precursors to phosphazenes via elimination of Me_3SiCl. Preliminary evidence indicates that the thermal elimination does indeed occur in some cases. The N-silylphosphinimines are also potential precursors to another type of novel Si-N-P compound, i.e. three-coordinate phosphoranes:

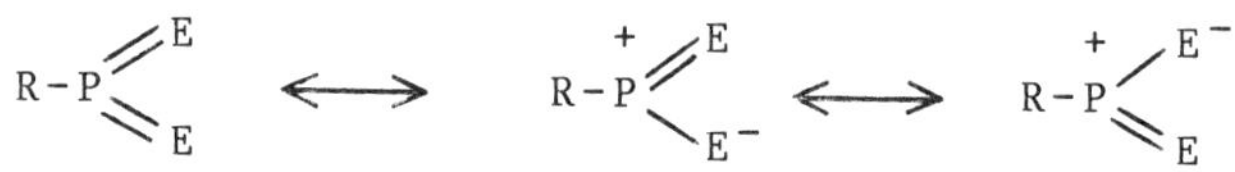

E = CRR', NR

Our interest in such compounds stems mainly from the possibility that they might be useful precursors to new classes of phosphorus-containing polymers or cyclic oligomers. Functional linkages such as E = $NSiMe_3$ or $CR'SiMe_3$ could serve as sites for condensation-polymerization reactions, leading to novel cyclic or polymeric systems,

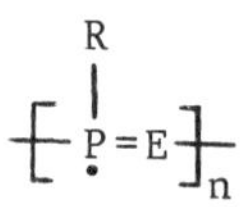

where E = N or CR'. These materials are isoelectronic with the electrically conducting polymer, poly(sulfur nitride), $(SN)_x$.

Silylaminophosphines, once again, are appropriate precursors to such low coordinate systems. Treatment of chloro substituted silylaminophosphines with Me_3SiN_3 at room temperature results in spectroscopically pure azido phosphines following removal of Me_3SiCl (eq 10). In the case where R is a mesityl group, heating the

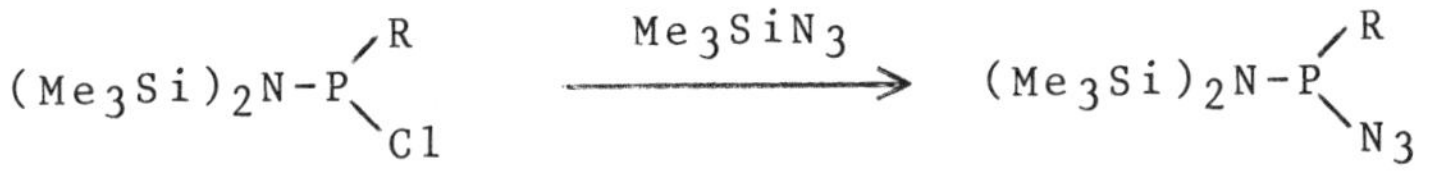

(10)

R = CH_2SiMe_3, Ph, Mes, $N(SiMe_3)_2$

azido phosphine produces a loss of N_2 and a subsequent [1,3] silyl shift. At 80°C the cis 4-membered ring system (^{31}P δ -19.7) is formed while heating at 175°C produces the more thermodynamically stable trans isomer (^{31}P δ -24.7) (eq 11). The two crystalline isomers

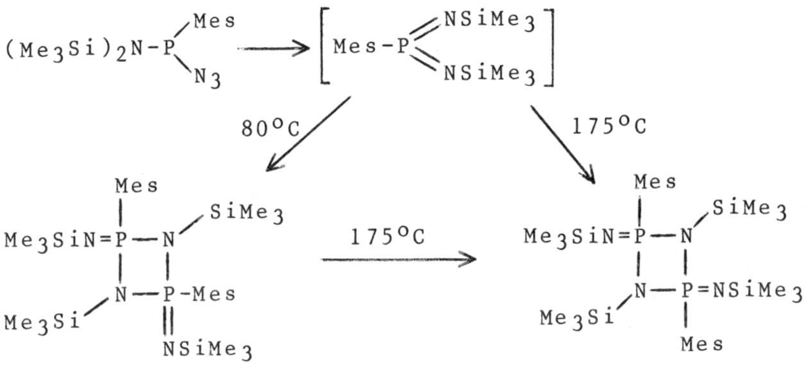

(11)

cis trans

were purified by recrystallization from hexane and were fully characterized by elemental analysis, cryoscopic molecular weight determination, and ^1H, ^{13}C, and ^{31}P NMR spectroscopy (13). These rings represent the dimeric forms of an intermediate three-coordinate phosphorane. A stable analog of this intermediate, $(Me_3Si)_2NP(=NSiMe_3)_2$, has been isolated by a similar sequence of reactions (14).

A stable three-coordinate phosphorane and two other novel ring systems were isolated from the reaction of CCl_4 with $(Me_3Si)_2NP(CH_2SiMe_3)Mes$ (eq 12). The reaction proceeds with elimination of both $CHCl_3$

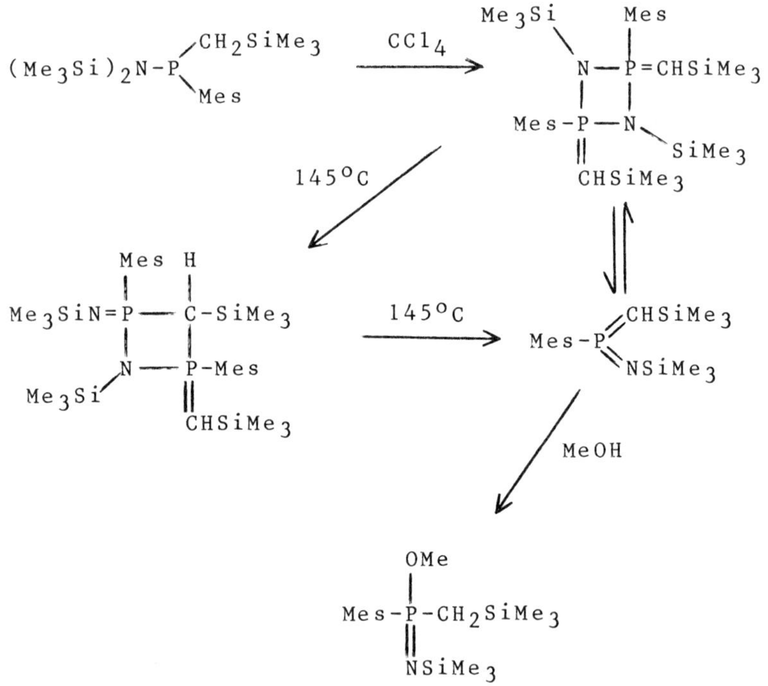

(12)

and Me_3SiCl and formation of a white crystalline solid which has been identified as the methylene substituted 4-membered P-N-P-N ring system (^{31}P δ +22.2). Upon distillation (b.p. 91-92°C/0.01 mm), the monomer $MesP(=NSiMe_3)(=CHSiMe_3)$ was collected as a colorless liquid (^{31}P δ+122.7). The monomer reverts to the dimer on standing at room temperature overnight. It readily adds methanol across the P=C bond to form the P-methoxy phosphinimine.

On heating at 145°C for two hours, both the monomer and dimer form a novel 4-membered P-N-P-C ring compound. The ^{31}P spectrum of this compound consists of two doublets of equal intensity at δ+10.9 and δ-10.6 with J_{PP} = 15.3 Hz. The downfield shift correlates with that of the phosphorus in the ring system containing $Me_3SiCH=$ substituents, while the upfield shift is similar to that of the 4-membered ring systems containing $Me_3SiN=$ substituents at phosphorus. The rings have also been characterized by ^{13}C and ^1H NMR, elemental analysis, and molecular weight measurements using both mass spectroscopy and cryoscopic techniques.

Preliminary studies indicate that the reactions of some of these dimers with $RPCl_2$ result in elimination of Me_3SiCl and ring opening reactions. Details of these investigations and complete characterization of these ring systems will be reported elsewhere.

Acknowledgments

The authors thank the U.S. Army Research Office, the U.S. Office of Naval Research, and the Robert A. Welch Foundation for generous financial support.

Literature Cited

1. Wisian-Neilson, P., Neilson, R.H. J. Am. Chem. Soc. 1980, 102, 2848.
2. Neilson, R.H.; Wisian-Neilson, P. Inorg. Chem. 1982, 21, 3568.
3. Wilburn, J.C.; Neilson, R. H. Inorg. Chem. 1977, 16, 2519.
4. Wilburn, J.C.; Neilson, R.H. Inorg. Chem. 1979, 18, 347.
5. Neilson, R.H.; Wisian-Neilson, P.; Wilburn, J.C. Inorg. Chem. 1980, 19, 413.
6. Wilburn, J.C.; Wisian-Neilson, P.; Neilson, R.H. Inorg. Chem. 1979, 18, 1429.
7. Wisian-Neilson, P.; Neilson, R.H. Inorg. Chem. 1980, 19, 1875.
8. Li, B.-L.; Engenito, J.S. Jr.; Neilson, R.H.; Wisian-Neilson, P. Inorg. Chem., in press.
9. Wisian-Neilson, P.; Neilson, R.H.; Cowley, A.H. Inorg. Chem. 1977, 16, 1460.
10. Appel, R. Angew. Chem. Int. Ed. Engl. 1975, 14, 801.
11. Appel, R.; Peters, J.; Schmitz, R. Z. Anorg. Allg. Chem. 1981, 18, 475.
12. Kolodiazhnyi, O.I. Tetrahedron Lett. 1980, 21, 3983.
13. Xie, Z.-M.; Neilson, R.H. Organometallics, in press.
14. Scherer, O.J.; Kuhn, N. Chem. Ber. 1974, 107, 2123.

RECEIVED March 30, 1983

INDEX

INDEX

Production by Paula Bérard
Indexing by Florence Edwards
Jacket design by Anne Bigler

Elements typeset by The Sheridan Press, Hanover, PA
Printed and bound by Maple Press Co., York, PA